53

And I Shall Fly

"The team"—Lin and Lewis Leigh at Toronto in the summer of 1957, shortly after Lewie's retirement from the RCAF.

Z. Lewis Leigh

And I Shall Fly

CANAV Books

1st printing

Printed and bound in Canada by
T.H. Best Printing Company Limited, Toronto

Published by
CANAV Books
Larry Milberry, Publisher
51 Balsam Avenue
Toronto M4E 3B6
Canada

Canadian Cataloguing in Publication Data
Leigh, Z.L. (Zebulon Lewis), 1906-
And I shall fly

Includes index.
ISBN 0-9690703-6-5

1. Leigh, Z.L. (Zebulon Lewis), 1906- . 2. Air pilots—Canada—Biography. 3. Air pilots, Military—Canada—Biography. 4. Aeronautics—Canada—History. I. Title.

TL540.L44A3 1985 629.13'0924 C85-098513-7

To my wife, Lillian Jane, known to all our friends as "Lin," for her quiet courage and support through the years, which made it all possible.

"Lin" died on September 6, 1984, after a year-long battle with cancer. She maintained her usual great courage right to the end.

Z.L.L., 1984

Contents

The photographs appear between pages 116 and 117.

Preface

Twice in the past suggestions were made by good writers that they undertake to write this story, but I was not particularly interested in having it done. Later on, however, I decided to write it myself, and completed the work with the aid of my flying log books, a fairly complete pictorial record with newspaper clippings, etc. and my memory. Although I had managed to record the story it was evident that it needed the attention of a good author. Flight-Lieutenant Hugh A. Halliday, a retired RCAF historian and now well known aviation writer and Curator of War Art at the Canadian War Museum, came to my rescue. He has extensively revised the book, making it a far better piece of work.

Some may feel that we have shown a bit too much detail of flying times between various places but this was done deliberately as the comparatively slow speeds we achieved in the old airplanes are being forgotten now in the light of modern jet speeds. A good many old pilots and air engineers are mentioned in connection with various flights and air bases. They are all good friends of mine and it was our aim to be as accurate as possible in order that they, when reading the book, would agree that what has been written is correct.

We have tried to show the effects of the Great Depression of the early 1930s on aviaiton, with its terrible economic problems and the difficulty of staying in the business. What has been described for me was also the situation of many of my pilot and engineer friends. A deep resolve and dedication were necessary to stay in aviation.

Bush flying of the 1930s is covered, with its lack of radio, weather reports and rescue facilities. The beginning of airline operations in Canada is described in the early days of Trans-Canada Air Lines, and then later on in World War II the commencement of the RCAF Air Transport Command, now a very famous organization. In general we hope the book will portray a good cross-section of Canadian aviation from the late 1920s to the late 1950s.

My nephew, Kenneth Leigh, provided good suggestions and also the title of the book *And I Shall Fly.* Mr Allen Martin of Air Canada got me started with the first writing. Mr. J.D. Hendricks of Grimsby, formerly of the RCAF, and Mrs. William Wood, whose husband was also in the RCAF, provided the equipment and the typing of the first writing. To all these people we say thank you.

Z.L.L.
H.A.H.

First writing 1969
Second writing 1972-3

And I Shall Fly

by

Group Captain Z. Lewis Leigh (RCAF, Retired)
Order of the British Empire (Military Division)
Efficiency Decoration
Legion of Merit—Degree of Officer (United States)
McKee Trans-Canada Trophy, 1946
Order of St. Lazarus of Jerusalem
Companion of the Order of Icarus
Member, Canada's Aviation Hall of Fame
Companion of the Order of Flight (City of Edmonton)
Brotherhood of the Silver Wings (Government of the Northwest Territories)
Member of the Order of Polaris (Government of Yukon Territory)

1

The Adventure Begins

Our tortured engines, running at full throttle, pulled us through the solid cloud and darkness as we carried out the whole instrument landing procedure again, dragging our broken undercarriage and with our flaps fully down because of the damaged hydraulic system. It was blind flying at its blindest. The radio crackled as Vancouver ground control advised us that fire engines and an ambulance were standing by, so we began our descent again. This time, I knew, would be our final approach, one way or another; the Lockheed could not go around another time.

The descent continued. We put our landing lights on and I told the co-pilot to be prepared to chop both engines as soon as I gave the order.

Suddenly the ground was rushing at us—we were heading right into it. I jerked back on the control column and as she levelled out I yelled "Cut." The co-pilot banged the master switch but the props were still ticking over as we hit the ground.

The aircraft had touched down in a near-perfect belly landing. It slithered across the field, turning slightly as I sat helpless at the controls. As it stopped the passengers and the stewardess scrambled out, followed by the co-pilot and myself. We retreated into the murk, but luck was with us—there was no fire.

It was January 26, 1940, and although I did not know it my life had reached a decisive point. For the past 11 years I had been deeply involved in civil air transport work. For the next 17 I would be largely concerned with building up the air transport branch of the RCAF. The near-disaster that January night might well have put a "period" to my life; instead it was a hyphen that joined yet separated two parts of it.

I was born in Macclesfield, Cheshire, England, on June 19, 1906, the first of five children (one girl, four boys). I was christened Zebulon Lewis Leigh. The Zebulon was a Biblical name which had been in the family for generations, but throughout my life I have been called Lewis (''Lewie'' to friends).

My father, a skilled house-builder and cabinet-maker, was deeply involved in the British Territorial Army. My mother was a good-hearted, considerate woman, proud of her family and proud of her home.

Father, something of an adventurer, took it into his head to emigrate to North America. The family, which then consisted of my parents, myself and my sister Ida, came over in 1909. My brothers Bert, Dick and Eric were all born in Canada, we having settled in Lethbridge, Alberta.

My father returned overseas during the First World War as a member of the 39th Battery, Canadian Field Artillery. He spent three years in France, rising to the rank of sergeant. He was almost a stranger when he came home, but he quickly set about getting to know us again. The boys he would take target shooting, and his stories about life in the Artillery fascinated us. Father's pride in his country and his service affected my life to no small degree.

A postwar slump in the house-building industry imposed some financial hardship on our family and I left school at the age of 14. My education since then has been almost entirely self-instruction or courses taken in conjunction with civil or military flying duties.

My first permanent job was delivering groceries, first with a horse and wagon, a little later with a four cylinder McLaughlin-Buick truck. Then I drove a truck for the Hudson's Bay Company. Finally, after passing civil service exams, I was accepted as a postal clerk in Lethbridge.

During the period of Post Office service I joined the 20th Field Battery, Canadian Artillery, a militia unit stationed at Lethbridge. I rose from gunner to sergeant, and then was commissioned as a lieutenant. Each year our battery attended summer camp at Sarcee, near Calgary, where we competed with other militia batteries under the supervision of officers of the Royal Canadian Horse Artillery, Regular Force. Our unit was particularly efficient, for on one occasion we won the Gzowski Trophy given to militia batteries. Indeed we were sometimes a bit *too* efficient. Standard target practice was to fire at wooden tanks pulled at various speeds on the range, and I caught hell one day when my section blasted a stationary target to smithereens.

During one of these camps we were to take part in a general march past where the salute was to be taken by Gen. Sir J.H. MacBrien, Chief of the General Staff. On the day of the march past my rather quiet horse was taken away and I was given a very spirited one. The horse had been used by a Regular Force battery commander and was accustomed to a position right up in front of the battery. My position as a very junior officer was in the rear of the unit.

Our battery was following immediately behind ''C'' Battery, RCHA, commanded by Lieut.-Col. Stockwell, a unit which had provided our instructors both in Lethbridge and Sarcee. Once the march past began my horse, whose dignity far exceeded mine, decided to get to the forward end of things. I tried desperately and in vain to rein him in, and finally found myself almost alongside Col. Stockwell. Just as we neared the saluting base on which Gen. MacBrien stood, my horse shied and jumped sideways, catapulting me from the saddle. I landed, spread-eagled, in front of the saluting base. Col. Stockwell's voice boomed out, "Leigh, who ordered you to dismount?"

Needless to say, I got out of the way—on foot. By the time my horse was caught the march past was finished. I have never forgotten the mortification of that moment.

On occasion we would see an RCAF aircraft flying over us doing practice spotting for the guns. Sitting in an aircraft looked much better to me than sitting astride an artillery horse. Besides I was beginning to soak up the lore of Canadian pilots in the Great War, so I decided to get into the flying business. The thought of sitting in an open cockpit, behind a good aero engine, controlling my own flight through the air, became an overwhelming desire. I wrote a letter to the Air Board in Ottawa asking to be enrolled in the Air Force but was politely turned down. It became clear that if I were to fly I would have to make my own arrangements.

Aviation in general at this time consisted mainly of small military air forces, usually flying open-cockpit 110-125 mph biplanes. Civil aviation was struggling hard to make an appearance, with the beginnings of semi-regular air services, particularly between Britain and Europe, and also in some parts of the United States. Some large, awkward aircraft were appearing which carried a few passengers at 100-110 mph. Bush flying operations were just starting in Canada in the form of forest fire fighting and patrols, with special flights to outlying points when required. Trans-atlantic and other long-distance flights were being tried, many unsuccessfully, very much on a daredevil basis. Other than that, flying was looked upon as a sport rather than a business. Flying schools were springing up and barnstorming operations were thriving. These were irregular, small-scale affairs of one or more pilots taking their airplanes, usually surplus first World War training planes with speeds of 70-90 mph and costing one to two thousand dollars, to small towns across the country. Usually they would put on an aerobatic display over the town selected, then land in a nearby open field and begin the business of carrying the townspeople for joy-rides.

A valuable counsellor at this time was J.H. "Jock" Palmer, who built and sold radios in Lethbridge. He had flown in the RAF in 1917-18 and owned a Curtiss JN-4 in which he sporadically barnstormed. His reminiscences spurred me on to flying, though he never took me up. More important, he advised me on how to break into flying. It was at his suggestion that I had written to the Air Board, and when my application was refused he advised me further.

Charles B. Elliot of Lethbridge and George Ross, a rancher from south of the city, teamed up and purchased a de Havilland Gipsy Moth early in 1929, followed by a second later in the year. They formed Lethbridge's first flying school and, when pupils were few, barnstormed around southern Alberta and Saskatchewan. The company was called Southern Alberta Airlines and its first flying instructor was Joe Patton of Calgary, who had been taught to fly by Jock Palmer and F.R. McCall of Calgary. In time Patton was joined by Bob Kern of Taber.

I enrolled immediately. The charge was $30 an hour for dual instruction. The field itself was just a patch of open prairie southeast of the city. It had one or two hollows that one avoided on landing, and gophers. A T-shaped metal hangar accommodated the aircraft, which was backed into it when necessary. In later years a proper airport was laid out on the west side of Lethbridge and named Kenyon Field, in honour of A/C Herbert Hollick-Kenyon, a great pilot and a good friend of mine.

Charlie Elliott ran an automobile taxi service in addition to his flying company. He had not yet learned to fly so we both started training under Joe Patton. Lethbridge is notorious for strong winds, and as our Moths were very light we often took our flying instruction very early in the morning, usually just after dawn, while the morning sun tinted the snows on the peaks of the Rockies to the west. If the winds permitted we might also go up in the evening.

I was still working for the Post Office so my schedule was fairly tight. As a rule Charlie Elliott would pick me up in one of his taxis en route to the airfield a couple of miles away. Sometimes this wasn't possible and therefore, not having a car of my own, I had to walk to the field.

Charlie and I were given a very thorough, though short course by Joe Patton, with emphasis on spinning and spin recovery. We both soloed with a very minimum of hours (I cannot recall the exact number of dual hours involved as my flying log books and licences were destroyed by fire at Camp Borden in 1932). Charlie made a good solo landing, but I had trouble judging height and damaged the undercarriage.

At this time the company purchased a Curtiss Robin monoplane with a Curtiss Challenger 180 hp engine. This was considered a very fine airplane in its day, used mainly for short-haul passenger trips and barnstorming.

I had concluded early in 1929 that flying was going to be my way of life so I started to build up solo time. This was not easy as I did not have much money to spare. I helped where I could on barnstorming trips, going along to swing propellers, sell tickets and strap passengers into their seats. For this I was allowed to fly the aircraft to and from various places without any charge for flying time. When I *was* buying solo time by the hour, I spent much of it practising loops, rolls and spins, until I felt reasonably comfortable with these manoeuvres. I was also getting more expert at selecting places to land away from the home base, judging the lengths of various fields, type of surface, ditches and so on. This knowledge was vital in barnstorming and later in bush flying.

I had decided not to bother obtaining a private pilot's licence but to concentrate all my efforts on getting the commercial licence. This required a minimum of 50 hours solo time prior to test, plus an examination in airmanship, navigation, etc. I also decided to try to obtain an air engineer's licence in "A" and "C" categories, which would permit me to take an airplane away barnstorming and remain out for a few days by inspecting my aircraft and signing for its airworthiness each day.

Eventually the day arrived when Inspector Howard Ingram of the Civil Aviation Branch, Department of National Defence, flew to Lethbridge in his Moth. It was agreed that I would take my commercial flying test and a little later on to try for an air engineer's licence. I was a bit tense during my flying test but managed to land near enough to the marker that Ingram had set up on the field. I also carried out the rest of the operation to his satisfaction. Later on I wrote the air engineer's examination. Both tests I passed. (The licences were both destroyed by fire at Camp Borden on January 11, 1932, and then replaced by the Department of National Defence.)

I could now legally carry passengers for hire so Charlie let me carry out some weekend barnstorming trips on my own and also to instruct in the flying school on calm mornings and evenings. At this time the holder of a commercial licence could give flying instruction without having a special instructor's licence. One of the people instructed was George Ross, part owner of the company. George had been a wartime pilot so he only required refresher training to become active again. He was a good pilot and very enthusiastic about the future of aviation.

In the long run, instructing was the most stable revenue-producer for Southern Alberta Airlines, but the barnstorming was great fun, although it was becoming less lucrative as the airplane ceased to be a novelty to the public. Barnstorming had about it a touch of the gipsy life, when you could navigate by following the "iron compass" (railway tracks) and find out where you were simply by buzzing a hamlet and reading its name on the grain elevators. It was not uncommon to cross the path of barnstormers operating out of Calgary or Moose Jaw, and practically all the pilots in western Canada knew each other.

With southern Alberta's frequent strong winds, we had to learn to land under very gusty conditions. Usually we flew over a field while helpers got into position on the ground, then tried to land as close to them as possible so they could grab the wingtips and hang on. Otherwise the Moth or Robin might blow over. It was always in the back of our minds that in an emergency, with a high wind blowing, we could try to land in the lee of a haystack. Even if we overshot, we would do less damage by nosing slowly into the haystack than by allowing the winds to flip the plane over.

The dangers associated with winds were underlined one day when a friend, Ernie Boffa, arrived from the US piloting a Stinson high-wing monoplane. The wind was gusting and the helpers could not get to him fast enough. The Stinson flipped onto its back and it was quite a job to get it upright again and make it flyable.

In the late summer of 1930 Charlie Elliott and George Ross decided to expand by opening another flying school and barnstorming service at Medicine Hat, about 100 miles east of Lethbridge. They offered me the job of operating it as combined manager, air engineer and pilot-instructor, all for the then-reasonable sum of $175 a month. I immediately resigned from the Post Office Department—I could now devote myself full time to aviation.

2

At The Hat

It was autumn 1930 before arrangements were completed for the use of a small wooden hangar on the Medicine Hat airfield. The newest of our two Gipsy Moths, CF-AGJ, was allocated to me to start the new operation. I loaded my baggage into the front cockpit, tied it down, then took off for my new domain.

The Medicine Hat airfield was a flat expanse of prairie encircled by the yellow metal cones of the boundary lights. There was one hangar, just large enough for one airplane, a field floodlight for night landings, a structure that housed the fuel pumps, and a Department of Transport meteorological building, run by Charlie Pickering.

Medicine Hat was a point on the Prairie Airmail that was flown between Winnipeg and Calgary by Western Canada Airways. Its Boeing 40B-4s, Fokker F-14s, Laird LC-B 200s and a Consolidated 20 Fleetster, were frequent visitors at "The Hat," and their crews were a veritable Who's Who of Western flying—Herbert Hollick-Kenyon, Milton Ashton, Con Farrell, Jock Jarvis and others. Periodically air engineer Sammy Tomlinson would show up when an airplane needed extensive repairs. Presiding over this bustle was Bob Stevenson, who managed the airport, sold fuel and loaded mail; he never stopped to count all the hats he wore.

And into it all came myself—a new pilot occupying a small shack of a hangar. My workshop was one corner of the hangar, heated by a natural gas element that burned inside a 45-gallon fuel drum with a hole cut in the top. Assisting me was young Foster "Fuzzy" Clenell, who was breaking into aviation the same way I had—by swinging the propeller, refuelling the plane, booking passengers and sometimes coming along with me on barnstorming excursions.

One day I was standing by the hangar when a white Moth came in for a landing. It bounced a bit on touchdown, then taxied up to the building. It was CF-AAA, the property of the Canadian Flying Clubs Association. Out clambered the president of the association, Gen. J.H. MacBrien (retired), who had only recently learned to fly—and hadn't yet, by my observation, learned to land too well.

The General looked at me closely, then remarked that my face was familiar. I reminded him of the incident some years before when I had been trying to catch a horse in front of his reviewing stand. Once we had had a good laugh at that story he admitted that his own landing had not been the best. I went round with him on a couple of circuits and helped to iron out his touchdowns, after which he left. I never saw the General again, though I was to hear from him in later years when my flying work led me into cooperation with the RCMP, of which he was by then the Commissioner.

Initially I was very busy at "The Hat," giving flying instruction, taking up passengers for joy-rides and barnstorming at local fairs. I stayed in a hotel about two miles from the field and dined at restaurants. My daily routine was pretty spartan—up before dawn, walk to the airport, roll out the Moth, inspect and test fly it and then be ready for my first pupil of the day.

Barnstorming was always good for a few dollars more as well. Sometimes I would fly out to a town where a fair was being held; on other occasions I would just select a spot that no barnstormer had approached for some time, go out and "shoot up" the place, then land in a nearby field and wait for the paying joy-riders to come. When airplanes were a novelty, the arrival of a barnstormer was cause enough for the whole town to close up shop.

There was one person in Medicine Hat who took an extraordinary interest in flying. Lillian Jane Bowker arrived one day at the field with a party of friends, all of whom wanted to be taken up. Everyone was content with a short flight around the city—everyone, that is, except Lillian, who insisted that I do a loop and a roll and who thoroughly enjoyed these aerobatics. From then on we began to see more of each other and to fly fairly regularly. "Lin," as I shall call here hereafter, flew often with me from then on; through more than 40 years of marriage she was partner and helpmate through hell, high water and hard times.

And hard times were soon to break upon us. When I began the Medicine Hat operation there were plenty of people wanting flying instruction, the bread and butter of my business, but by late 1930 all business—flying and otherwise—was on the skids. It became apparent first at the flying school, where fewer and fewer people showed up for training. Then it spread to the small towns, now growing grimy with the dust storms of the Dirty Thirties, where it became harder for people to find five or ten dollars for a ride. Then the problem became a personal one; Southern Alberta Airlines sold off my Moth. I was on my own without an airplane.

To the rescue came an old friend from Lethbridge days, Ernie Boffa. Ernie owned a Waco 9 biplane powered by a Curtiss OX-5, a liquid-cooled eight-cylinder design that dated to the First World War. CF-AOI was a good machine, though a bit underpowered. Ernie checked me out on it and I subsequently bought an interest in it. It was our idea to pursue the flying business, despite the recession, and to base the Waco in "The Hat."

Though we still offered flying instruction to all who were interested, our chief concern was barnstorming. To add colour to this Ernie introduced a wing-type exhibition parachute. The chute itself was packed in a long canvas bag and was tied to a wing strut, on a fore-and-aft axis. The jumper wore a leather harness connected by rope to the parachute in the bag. Once in the air he would climb out on the wing, then dive off, trying to avoid being hit by the tailplane as he did so. The rope would drag the parachute out of its bag and the jumper was on his way. Crude as it was, it worked, and it proved to be a great crowd pleaser when the ordinary repertoire of loops, spins and rolls grew stale.

Then Ernie and I decided to *really* create a stir. Our idea, concocted appropriately in a beer parlour, was for our Waco to fly over a crowd towing a daredevil hanging by his teeth to a rope that dangled from the undercarriage. At least, that was the way the spectators would see it, and that was how we advertised our plan.

Actually we were cheating a little. The rope, 12 feet long, was tied to the undercarriage. About every foot along it was a large knot and at the end of the rope there was a steel snap hooked to his harness. Ernie was to climb out of the front cockpit onto the wing while I flew low over the crowd. From there he was to reach the undercarriage and climb down the rope, then hang onto that last knot with his hands near his mouth, thus simulating the "hanging by his teeth." I would circle the crowd, after which he would climb the rope again and return to the cockpit.

For this stunt Ernie was not to wear a parachute. We posted some friends in the middle of the field with the understanding that if anything went wrong they were to wave to me, for once Ernie climbed off the wing he would be out of my sight. The problem was that though they could signal that something was amiss they could not describe exactly *what* was wrong. In our beer parlour schemes we never considered this point.

A substantial crowd had gathered on and around the Medicine Hat airfield when we took off. I climbed to 300 feet and then flew over the onlookers, keeping the aircraft in a wide circling turn. Ernie crawled out onto the wing, then grimly let himself down over the leading edge, heading for the undercarriage. His body created wind resistance, so I opened the throttle more. The underpowered Waco was fighting to maintain altitude. I waited for a minute or two, giving Ernie time to ease himself down the rope, then began my circuits over the crowd. As I completed the routine I glanced towards the centre of the airfield. There were our friends, waving frantically. Something was wrong—but what?

Exhilaration turned to sobriety. I decided to do one more pass over the crowd and see what happened. The waving became more frenzied. I guessed, rightly, that for some reason Ernie could not climb back up the rope and that he was still dangling there. My job was to try to get the airplane down without killing him.

For a start I selected a smooth, grassy part of the field, far away from the crowd. I chose to come in faster than usual to keep Ernie "streaming out" behind, the flatter the better. Then, too, I figured I should keep the tail up as long as possible, lest the iron tail-skid shoe bash him. At that time three-point landings were normal; of necessity this must be a wheel landing.

Down I touched, then rolled. I could visualize Ernie being dragged along behind at 40 miles an hour, then 30. How was he? As I came to a halt I sat there frozen with fear. "I must have killed him!" I thought. From under the plane crawled a figure in white flying overalls, dirty, somewhat bloodied about the knees, elbows and nose. Our eyes met, and without a word, as sober as judges, we shook hands. At that moment our friends arrived, Lin among them. She looked from Ernie to me and back again, then uttered the most appropriate words for the occasion: "Come to my house and have a drink. You both need it."

I learned afterwards that Ernie had been unable to climb back again because of the slipstream, aggravated by the extra power I had applied to maintain altitude. Snapped to the rope, he could not fall, but he was whipped around in the slipstream. As I settled down he was pumping his legs like crazy, bounding along in giant 30-foot strides behind the Waco. In the crowd there was pandemonium; more than one woman fainted.

Subsequently I received a ticking off from Inspector Ingram of the Civil Aviation Division, Department of National Defence. I promised never to try a similar stunt again.

Mad behaviour among barnstormers was not uncommon. Charlie Elliott, who had learned to fly with me in Lethbridge, went hunting for antelope around Nemiskam using a Moth, and armed with a rifle. Having selected one animal, he buzzed in pursuit, taking aim while he held the stick between his knees. He picked off his target, but a dune got in his way and Charlie cracked up. He was barely scratched, but the Moth was wrecked. Minutes later a game warden arrived and confiscated the antelope. Charlie had to phone me to fly down and pick him up.

On another occasion, while barnstorming near Vauxhall, Alberta, in his Robin, CF-AHF, Charlie damaged a landing gear fitting. Determined to do his own repairs, he borrowed a welding outfit from a local garage and set to work. In the process he set the plane alight and 'AHF went up in flames.

One very soggy day I received a phone call from Maple Creek, Saskatchewan. A woman was in urgent need of an appendectomy and her doctor wanted her flown to Medicine Hat. With a friend, Fergie Fraser, in the front cockpit, together with one or two blankets, I set out on the mercy flight, hedge-hopping through the rain. Near Maple Creek the rain reached cloudburst proportions. The old OX-5 could not take it. We could not go back, so I picked some open ground, a stubble field, and landed. As soon as I was down the Waco became mired in the muck.

We left the plane and found a farm house. The occupant, an immigrant who spoke little English, was hospitable. I phoned into town where the doctor reported improvement by his patient. The flight to Medicine Hat, he informed us, was no longer necessary. Fergie and I scrounged a meal and bunks at the farm that night. The next day, with the help of the farmer's hired hands, we got our plane onto more solid ground, faced it down a slope and staggered into the air. We arrived safely back at Medicine Hat, tired and dirty, having accomplished exactly nothing.

Fergie Fraser himself merits some mention. He was a fireman by trade, a boxer (Alberta amateur lightweight champion) and a fine fellow. He took flying lessons from me and in turn gave me boxing lessons. We set up a mat in the hangar, where he gave me many a rough time. It seemed at the time I spent most of my lessons picking myself up off the mat, but I learned, and what I learned once saved my life.

Fergie had picked up a Model T Ford open-back truck, 1917 or 1918 vintage. It had a brass radiator and needed to be cranked. The thing cost $25, half of which I provided, so half of its time was spent transporting us back and forth from the airfield. We also used it to call on our lady friends. The neighbours for blocks around were always aware of our social calls as the Ford had long since lost its muffler and sounded much more like a battery of field artillery in action.

I had developed a very violent temper which tended to get me into trouble—and indeed, it still does. On one occasion I had taken the Waco off, solo, to barnstorm in Saskatchewan. Our OX-5 engines had a habit of tossing off their overhead valve rocker arms and push rods. The rocker bearings were poor and tended to wear rapidly, with the result that they would come apart violently. The parts would go sailing back over the pilot's head and the cylinder from which they came would go out of action. With one or two cylinders out the Waco became so underpowered that a forced landing was inevitable. We were so used to this kind of trouble that we normally carried a few spare rocker arms and push rods wherever we went.

On this day the weather was perfect for barnstorming and everything seemed to be going well when suddenly, *sprong,* the usual valve-action parts went streaking over my head and the engine became rough and feeble.

In fury and frustration I pulled the wooden control stick from its socket (easily removable) and violently beat the small windshield in front of me into atoms. This, of course, solved nothing; I was still losing altitude. I calmed down enough to replace the stick, make a forced landing and repair the engine. Once I took off again, though, I sorely missed that windshield. The slipstream full in my face was retributive justice and it was weeks before I could get a replacement windshield.

In January 1931 Ernie Boffa contacted me. He wanted to move the Waco from Medicine Hat, but he knew where a replacement machine could be obtained fairly cheaply. It was agreed that he would take the Waco and arrange for the delivery of another aircraft to "The Hat." The plane which arrived in February, and which I purchased outright with borrowed funds, was a Command-Aire 3C3, CF-APQ. It had a front cockpit for two passengers and a pilot's cockpit. It was a rugged little biplane, but with an OX-5 engine it was, like the Waco, underpowered.

The Command-Aire had been built in Little Rock, Arkansas, and the makers offered $500 to anyone who could make more than half a turn in a spin with it. I tried many times to force it to spin, but it would not, such was its stability.

Although the Command-Aire was well built, I managed to overstress it shortly after I had acquired it. One afternoon I was carrying two hefty passengers who asked me to do a loop. I managed to complete the loop all right, but then I noticed that the controls were a bit queer. The aircraft seemed to be sloppy to handle. Looking up I saw the upper wings moving sideways a bit, back and forth. This was a shock. I started easing it towards the ground, turning as gently as I could. Eventually I landed with a sigh of relief. Once the two customers had departed I examined the aircraft and found that all the front centre section struts had pulled away from the fuselage. The welds had broken and we had been very fortunate to get back safely. I managed to get the local welding expert to reweld the struts, but never again did I do aerobatics with two heavy people in that airplane.

As tough as it was, the Command-Aire was the object of another of my fits of temper. With the instructing business going downhill every week, I decided to fly to Lethbridge with Lin and have her meet my family. On the morning of my departure my man Friday, Fuzzy Clennel, wheeled out the newly washed plane, with the engine warmed up and switched off, chocks in place, ready for departure. Lin arrived on time and was strapped into the front cockpit. I was anxious to get away; the wind was rising and a dust storm was taking shape in the west, where we were heading.

I climbed in, switched on and Fuzzy swung the propeller. The OX-5, however, was the most temperamental piece of machinery in creation, and it refused to start. Fuzzy kept swinging the prop while my patience evaporated. Finally I told him to take my place in the cockpit while I, encumbered with goggles, helmet, leather jacket, boots and breeches, angrily heaved and huffed at the stubborn prop. I swung it until I was blue in the face, and then my temper flared. Slowly, deliberately, I walked over to the lower left wing and drove my foot up through ribs and linen, tearing a good-sized hole in both the upper and lower wing surfaces. That ended any possibility of an early take-off.

We helped Lin out of the cockpit, then wheeled the plane into the hangar for repairs. Lin very nicely overlooked my display of petulance and said she would wait while we got things back in order. Meanwhile the wind was rising. We temporarily splinted the broken ribs and sewed up the linen fabric coverings, doped over the lines of sewing and allowed the dope to dry. By now my temper was cooling down.

Finally we were ready to try again. We wheeled the aircraft out, got Lin aboard and chocked the wheels. Fuzzy turned the engine over to suck in the mixture, I turned the switch on, Fuzzy swung the propeller once and the engine broke into life as sweetly as it had ever done. There wasn't a damned thing wrong with it! I warmed it up and we took off into a strong, gusty wind.

I battled that wind all the way to Lethbridge, 90 miles away. Normally the Command-Aire crused at 80-85 mph but with that head wind the trip took two hours, while the dust clouds churned below us, brown topsoil and men's dreams whipping away. The gusts rocked the plane as we descended, and when we landed we needed helpers to grasp the wingtips and keep her from blowing over before she was tied down.

It was a strange way for my parents to meet my best girl, but it was well worth it. They fell in love with Lin immediately and she reciprocated. We were feeling very rosy when we flew back to "The Hat."

By the spring of 1931 it was clear that my operation in Medicine Hat would go broke. Then a letter arrived from Sydney, Nova Scotia, from Maritime and Newfoundland Airways, offering me a job as a pilot on the East Coast. The aircraft would be float-equipped Fokker Standard Universals. I had never flown Fokkers and I had never worked on floats, but jobs were scarce. The salary was no great shakes either, but I accepted and informed them I would be there by the end of June.

First of all, though, there was some business to settle. I sold the Command-Aire to Maynard Patterson of Medicine Hat, together with all my tools and spares. Maynard had been a First World War pilot and had joined my flying school in 1930. Once I had been paid the best price for all these goods I found I was still a thousand dollars in debt—and those were 1931 dollars.

Since I was leaving the prairies, I asked Lin to marry me, debts and all. She accepted, and since there wasn't much time before our departure for the East, we flew the Command-Aire for the last time to Lethbridge. We were married on June 11, 1931, in St. Mary's Anglican Church in the presence of a small group of family and friends. We then flew back to "The Hat" and the next day boarded an eastbound train. Thus ended a colourful but financially strained chapter of my career.

3

Maritime Interlude

We left the train for a 24-hour stop-over at Sioux Lookout, Ontario, where I was to receive a quick checkout on float landings and take-offs. After taking a room at a local hotel we walked down to the Canadian Airways offices and docks on the lake shore. They were expecting me and had arranged to have a float-equipped Fokker Super Universal ready to go at the dock. I paid dearly from my limited finances for three hours of instruction. My instructor was Stewart McRorie, a good bush pilot of whom I would see a lot in coming years.

It was quite an undertaking to jump from small, open-cockpit biplanes to a Fokker Super, which in those days was classed as a huge aircraft. It had an enclosed cockpit seating two pilots side by side. Stewart was a patient instructor, and at the end of my paid time I was beginning to get the hang of float operations. My landings and take-offs were sloppy, but the principle of getting the aircraft ''on the step'' for take-off and the wheel landing position (rather than the three-point landing attitude) were becoming clearer. I could have used a week of intensive float operations to master the new techniques, but time and money would not permit.

On arrival in Sydney we were met by Robert McGowan, the manager of Maritime and Newfoundland Airways, and taken to a hotel before finding a more permanent place to live. Bobby McGowan outlined the company and its operations. They owned five Standard Universals, complete with floats, wheels, skis, spare engines and cases of spare parts, all of which had been purchased from the Dominion government. The equipment came from the RCAF's Hudson Strait Expedition and had been sold to the highest bidder as soon as the Expedition had completed its northern job.

Two of the aircraft, G-CAHE and G-CAHJ, were on their Hamilton floats, ready to fly. The rest were partially dismantled in the seaplane hangar. The hangar was on the harbour shore of North Sydney. It had been built during the war as a joint Canada-US project to provide air cover to convoys, and the HS-2L flying boats based there in 1918 had been commanded by Lieut. Richard E. Byrd. Despite the passage of time, the base retained excellent facilities with fine slipways.

Operations were spasmodic, consisting of barnstorming on floats around the Maritimes, special flights carrying business people and occasional flights to Newfoundland as well as St. Pierre and Miquelon, the French-owned islands. All in all, the business and financial picture of the company appeared as shaky as that of the company I had recently sold at a loss.

The staff included Charlie Roy (a wartime veteran and the only other pilot), engineer Morden Carter, who had come to the firm from Canadian Airways, and a young man named Donald Cameron who acted as general helper around the hangar.

Lin and I managed to rent a pleasant furnished apartment in Sydney. This meant, however, that I had to commute to the base, usually in Bobby McGowan's car.

I was soon familiarizing myself with the Standard Universal. It was smaller than the Super I had flown at Sioux Lookout. It had the usual Fokker plywood-covered wing and a small, single-seat, open cockpit forward of the wing and immediately behind the engine, a Wright J-4B of 200 hp. The passenger cabin had four seats and was separated from the cockpit by a wooden wall with a small peep door, about four by eight inches, through which the pilot could look every now and then to be sure that all was well in the cabin.

There were mechanical problems. The Hamilton floats had poor performance characteristics. In addition, although the floats were made of duralamin, the rivets used were aluminum, which corroded in sea water and frequently dropped out, causing the floats to leak.

The J-4B Whirlwind engine was a nine-cylinder radial with open push rods and rocker arms which had to be oiled frequently. Sometimes, due to wear or loosening of the locking nuts, a push rod would jump out in flight.

I taxied out on my solo familiarization flight and stayed on the water until I had the feel of the machine, and got used to its height above the water. This would be essential when the time came to land. I tried a couple of take-off runs, throttling back each time after I got the floats up on the step. On the third run I felt that I was as ready as I could be so I let her lift off. I climbed out for a few thousand feet, then tried some turns and a partial stall. I was getting the feel of the aircraft quickly. It was time to try my first solo water landing. I brought it in with some speed to spare and, when near the surface, groped gingerly for the water, trying to get the same height I had observed on take-off. The Fokker touched down with a couple of long, light skips, then settled in the water. With a surging sense of satisfaction I knew then that I could fly the machine.

From then on I was busy, flying businessmen from one point to another, or dropping into a Nova Scotia port to carry passengers on joy-rides. On one occasion I flew a press photographer to Saint John to get aerial shots of a huge dock fire. On most of these flights the company manager, Bobby McGowan, came along in the passenger cabin, helping out in any way he could once we had landed.

About this time (1931) a flight of various military and civil aircraft was organized to tour Canada: the Trans-Canada Air Pageant. It landed at a small airport near Sydney, so Lin and I went out to say hello to some of the pilots we knew. Included were RCAF Siskin fighters, the RCAF Ford Trimotor, and many commercial types, including a Pitcairn Autogyro (a predecessor of the helicopter) owned by the British Consuls Tobacco Company and flown by Don McPherson. As fixed-wing commercial pilots, most of us disliked it, possibly because it had to be flown differently from our normal aircraft. We felt it had little or no future. This dislike of helicopters stayed with me until I finally rode in one at Goose Bay, Labrador, many years later. However, I was never really happy riding in a machine which I was not capable of piloting.

Bobby McGowan had been trying for some time to organize an air mail service between Nova Scotia, Newfoundland and St. Pierre and Miquelon, hence the company's name, Maritime and Newfoundland Airways. He had a number of stickers printed to be glued to air mail letters advertising the service, but the Post Office withheld permission. Bobby decided to go ahead anyway with a passenger and air freight service, hoping that mail contracts would come through later.

As senior pilot, Charlie Roy undertook the first flight. The route was a straight line following the magnetic compass from Sydney to St. Pierre. This was a long water hop in the days before weather reports, radio, homing facilities or reliable engines. The return flight was to include a landing at Port aux Basques or Burgeo, Newfoundland. Charlie was gone a couple of days, and we were much relieved when he swooped back into Sydney.

Shortly after, Charlie left the company, leaving me the only pilot. Early in August I had to make the flight to St. Pierre and Newfoundland. We gave our aircraft, G-CAHE, a good check-over and found everything in reasonable condition with the exception of the floats, which were slowly leaking, as usual. The wing fuel tanks were topped right up. The fuel gauges were two boiler-type glass tubes mounted behind my head in the wooden wing section. The level of gasoline in the tubes indicated the percentage of fuel in the tanks. My compass was a Pioneer type, mounted on top of the nose cowling in front of my small windshield. The cabin was partly filled with parcels and so on for St. Pierre.

When we were ready to go, my engineer, Morden Carter, stood by the lines securing us to the dock. I had to stand up on the seat in my cockpit and place the starter handle in the dog which protruded from the top of the cowling in front of my compass. It was an inertia starter, which required a lot of heavy winding to get the starter flywheel turning over fast. When ready, I yanked the starter handle out and placed it in a bracket by my seat, then sat down and pulled the starter connection. The engine caught and began to run nicely. After a warm-up the lines were cast off and the engineer scrambled aboard. We then taxied out into the harbour. After turning into the wind I opened the throttle wide and quickly tested my magnetos by switching from one to the other. Everything was O.K., so I continued the take-off and then started to climb away. As I gained altitude I did a rough check of my compass by aligning the aircraft above a stretch of shore which approximated my course heading and then noted the compass reading. We were off to a good start.

Soon the shoreline was left behind. From then on the sound of the engine, the oil pressure readings and the fuel guages were very important, for a forced landing in the Atlantic below would have spelled disaster to our airplane and ourselves.

The good visibility held but the winds and white-caps below were increasing. It was essential not to drift too far south of my course as I might then miss St. Pierre itself. When I arrived overhead I could see the breakwater around the inner harbour but found that I must land outside due to the strong winds and wind direction, then taxi in through the harbour entrance.

The seas were heavier than I had anticipated. The landing was rough and I had a real battle on my hands to get in through the breakwater entrance. Once or twice I thought the machine was going to roll over, but finally we were through and in protected waters. People waved from dockside to show us where to tie up. Morden Carter and I were stiff and cold from the flight, there being no heaters in the open cockpit or the cabin, and we were relieved to be safely ashore.

The company had arranged with a local trading firm to be our agents in St. Pierre, so their representative looked after everything for us. The customs officers and gendarmes were all friendly and pleased to see us and our aircraft.

We were put up at a small hotel where we were wined and dined. That evening we were given a ride on one of the big, fast rum-running ships in harbour taking on their cargoes for the Rum Rows of Canada and the United States. The crews who manned the ships were an interesting lot —very respectable, not at all the rough types one would imagine. We returned ashore for a bit more evening drinking where we met many of the local people, including some of the ship-to-shore wireless operators who maintained contact with incoming and outbound vessels. Meeting them made us wish for wireless sets of our own. Though crude sets, operated with a key, had been installed in our Fokkers while they were in RCAF hands, these had long since been removed.

Next morning we were feeling none too healthy, but had to get under way. The agents had loaded our aircraft, including several large cans of rum. We were given two sets of bills of lading, one consigned to the "High Seas" and the other to the port of North Sydney. Thus, in the event that we were caught with the rum aboard, we could show the "High Seas" papers and say we had only put in temporarily on account of weather. I suspect these arrangements wouldn't have done us much good, and the authorities would have given us headaches no matter how many papers we had had.

The wireless operators informed us that the ship reports indicated fairly good weather, so we started the engine and taxied out. I had intended to avoid Newfoundland and go straight back to North Sydney. However, as we flew along the weather deteriorated and I was forced northwards to the Newfoundland coast, where I spotted a small town with some sheltered water. With the weather closing in fast I opted to land and taxied into the protected water.

Dories full of fishermen rowed out to meet us and guide us in. One dory with three men aboard rowed alongside and under the Fokker's beautiful burnished wooden wing. Standing up in my cockpit, I could hear their conversation but could not move quickly enough to stop them. Said one fisherman, looking up at the wing, "She's brass." The second one said, "No, she's copper." With that the third one took his oar and poked it straight up, knocking a hole in the lower side of the plywood covering. As he pulled the oar out he nonchalantly replied, "No, she's wood." I was barely able to contain my temper.

The fishermen showed us where to tie up. We had landed in Burgeo, and we were royally treated, as ours was the first airplane that most living there had seen. We took a room in one of the best houses in town. Later that evening a dance was organized in our honour. It was quite a sight to see fishermen in their jerseys and rubber boots dancing away. They were generous and open-hearted people.

Next day the weather gradually improved and by late afternoon it looked decent enough to start out. My engineer had made temporary repairs to the wing, so we taxied out and took off.

Once we were under way the weather began to close in again so I headed for the Cape North area, intending to skirt along the Nova Scotia coast in case I had to make a forced landing. The winds were getting stronger and my ground speed was decreasing. Finally I reached the coast and began following the shoreline, guided by the flashing lights of marine beacons. I was becoming concerned as the area did not look good for a forced landing (anyone who has toured the Cabot Trail will know what I mean). Moreover, my aircraft had neither landing nor navigation lights. I didn't even have a light in the cockpit for my instruments; in the inky darkness I checked them periodically with the aid of a flashlight I kept in my pocket.

Night had by now settled in properly, but I groped my way along, following the winking beacons. With an open cockpit and only a small windshield for protection it was miserably uncomfortable. On the other hand, it was much easier to fly and navigate at low altitude than in later years when the pilot sat in a closed cockpit surrounded by windows that were difficult to see through, particularly in heavy rain.

Eventually I saw the glow of Sydney. I swung low over the harbour, which was barely distinguishable by the light glow outlining it, and throttled down, feeling for the water I could not see. There were also ships in the harbour which I could not see, but I had to take a chance. The Fokker settled after a bounce or two and I taxied to a small dock at the bottom of the harbour. After tying up I went ashore and placed a telephone call. In short order our load was taken away and we were driven into Sydney, where a few drinks took the chill out of our bones and relieved the tension.

I learned later that the Revenue Service had been tipped off that we were coming and had been waiting for us. However, as the weather deteriorated so badly they had decided that we had been forced down somewhere and they had gone home. We eventually turned in our bills of lading which showed we had come to Sydney "in ballast." It was a bit too close for comfort, however, and we stayed with our normal passenger flying from there on.

About this time the North German Lloyd Steamships Company began catapulting seaplanes from the decks of their liners *Bremen* and *Europa.* The seaplanes were Heinkels, low-wing monoplanes with open tandem cockpits. They were powered by Pratt and Whitney engines driving three-bladed metal propellers. The crew consisted of a pilot and engineer. The seaplanes, loaded with important mail, were launched by steam catapults on the foredecks of the liners, usually while the vessels were about 900 miles from shore. On westbound trips the Heinkels landed at Sydney, refuelled, then carried out a night flight to New York City. The system speeded up mail deliveries by one or two days.

Early in November 1931, the seaplane from the Bremen landed in Sydney harbour just before dark. The crew, pilot Fritz Simon and engineer Rudolph Wucyknecht, tied up, refuelled and had dinner. Weather reports were favourable, except for fog in the Bay of Fundy. This satisfied the men, so they made a normal seaplane take-off and disappeared into the night.

Early next morning we were told that the seaplane had crashed in the Cobequid Bay area of Minas Basin, one of the upper branches of the Bay of Fundy. The crash had occurred in fog and darkness when Fundy's 60-foot tides were out, so that portions of the bay normally covered with water were then mud flats. The keeper of the Burncoat lighthouse heard the aircraft engine, the crash, and then shouts or cries, but was unable to get to them before the massive tides rolled in.

I took off from Sydney as soon as I could, carrying Bobby McGowan and an air engineer. We found the general area of the crash, but by now the tide was in full. We searched as long as our fuel would allow, then flew to Parrsboro, N.S., landed and tied up. This was the only place in the area where aviation gasoline was available. By the time we had refuelled and eaten, the tide had begun to ebb and our Fokker was sitting on the mud, far from the water. To coordinate our landings with tide conditions in the Parrsboro area had always been difficult. On one occasion we had had to force land when the tide was out. It had meant staggering through mud, pulling the seaplane ahead of the incoming tide, struggling to get it to shore. On that occasion the tide had begun to go out again when we had finished our engine repairs. These conditions can be appreciated only by those who have seen the Bay of Fundy, with its huge stretches of glistening mud when the tide is out.

In any case we carried on the search to the best of our ability, mostly in bad weather. The changing appearance of the bay and its inlets also made it difficult to keep track of those places that had been searched. By the third day we were all tired. Moreover I was suffering from a cold and sore throat brought on by frequent soakings and wet feet slogging about in tidal flows. At night we sought relief with one or two stiff drinks.

On the afternoon of the third day I spotted a body in the water. I flew to the nearest boat, which was a mile or two away, and by waggling my wings and flying in the direction of the body I got the boat to follow. I landed near the body, which was buoyed up by an air-filled aviation life preserver. We stood on the aircraft float and tried to lift the man out of the water but could not. The boat soon arrived and the people hoisted him onto their vessel. It turned out to be the pilot, Fritz Simon. He had been slightly injured but had died of exposure shortly before we found him. Further searches failed to turn up either the engineer or the main wreckage. Though bits and pieces of aircraft did float ashore, nothing more was ever found. I was later told that Simon had been a German count, but to my knowledge this was never confirmed. The crash ended the North German Lloyd catapult mail flights.

On return to base we required a few days to get our machine and ourselves back in shape after what had been a rugged operation. Awaiting me was a letter from Ottawa offering me a place on an RCAF blind flying and air navigation course which was to begin in January 1932. The course was provided for certain active commercial pilots and was to last three months. During that time we would wear uniform, either as officers (in the case of those already holding a commission) or as sergeants. I accepted the offer and arranged to have my existing commission as a lieutenant in the Field Artillery transferred to the RCAF. In the metamorphosis I thus became a flying officer.

The offer came at an opportune time, for my job was coming to an end. The Great Depression was taking its toll of small aviation companies and mine, Maritime and Newfoundland Airways, was preparing to fold up.

Lin and I said goodbye to our Nova Scotia friends, packed all our belongings into a huge metal trunk and boarded a train late in December 1931 bound for Toronto. There we stayed about a week while military tailors made me an RCAF uniform and supplied the necessary gear that went with it. We then took a train to Angus, Ontario, and from there we were driven to the air station at Camp Borden.

4

A Short Service Career

We arrived at Camp Borden on January 2, 1932, and found the station just beginning to recover from the New Year's celebrations. We were welcomed by the Orderly Officer, F/L A.C. "Briggy" Carr-Harris, who arranged for Lin and I to be given a meal in the Mess, had a room assigned to me on the second floor of the Mess, and then lined up transport to Barrie, where we found a room in a boarding house for Lin. She was the only wife present of pilots attending the course, because I was the only one of the group whose flying job had expired, so I had had to take my wife and trunkful of possessions with me.

Since there was not much spare space in Lin's room for the trunk, I made arrangements to store it in the basement of the Mess. This was to be the last time we ever saw it.

On checking in for the course, my commercial pilot's licence, air engineer's licence and log books were produced to show what experience I had had in the flying world, and were recorded by the RCAF, for which I was soon to be grateful. My civil flying time as shown by RCAF records and certified by S/L George E. Wait was 1114 hours. The types of aircraft flown were Gipsy Moth, Cirrus Moth, Waco OX-5, Command-Aire OX-5, Curtiss Robin OX-5, Curtiss Robin (Challenger), Kari-Keen, Puss Moth, Fokker Standard Universal, Fokker Super, Travel-Air and Stinson Junior.

On the course were many pilots who would stay with aviation throughout their careers. There were Charlie Gordon, Bill Page, Dennis Murphy, Ralph Milani, George Silke, Dinny Dinsmore, Gordon Steeves, Jeff Home-Hay, Bill Catton, Bill Dean, Milt Ashton, Frank Young, Jack Charleston and others. We spent a few days in ground school while the instructors found out how much we knew about ground subjects. Then, on January 8, 1932, flying began with checkouts to find out how good or bad our flying really was. RCAF instructors I encountered included S/L

George E. Brookes (later A/V/M), chief flying instructor; S/L Tommy Lawrence (later A/V/M), who gave me a progress test periodically; and F/Ls Roy Slemon (later A/M, Chief of the Air Staff) and F.J. Mawdesley (later G/C), who were then our senior navigation instructors. Flying instructors were F/Os Bill Lawson, Reg Soundy, Scotty Dougal and Boggy Glover. F/O Eddie Morris checked out my night flying.

At the station were others whose paths crossed with mine many times later on—W/C George M. Croil (later A/M, Chief of the Air Staff), then the Station Commander, plus P/O John L. Plant (later A/V/M and a number of times my boss and flying partner). John went with the advance party to the new air base at Trenton. P/O Frank R. Miller (later A/C/M and the first Chief of the Defence Forces) was there, as was F/O A. Dwight Ross (later A/C and winner of the George Cross) and F/L Jaggs Lewis (later G/C) who, in 1940, were my commanding officers in Numbers 5 and 11 (BR) Squadrons respectively. We flew Gipsy Moths, Puss Moths, Fleets, Avro Tutors and Fairchild 71s. The latter were used on group navigation flights.

On the morning of January 11, 1932, we had just arrived on foot at the hangar line and had "fallen in" by flights when the station fire alarm started to scream. In no time a column of smoke started to rise from the direction of the Officers' Mess. We were dismissed from parade and quickly ran down the road towards the Mess. It was on fire on the second storey near the end of the corridor where my room was. In fact, it was determined later that the fire actually started in the room across from mine.

With others I ran into the building and upstairs to try to save some of my belongings but found it was impossible to reach the room. The whole end of that section was well aflame. Since I could not reach my room I tried to help someone (I think it was Frank Miller) get some of his clothes out. The smoke was thick and the clothes in his closet had hanger hooks going in both directions over the closet rail. In trying to grab an armful we found they would not lift off, so we had to get out fast with little to show for our efforts. Some of the boys tried to save the Mess piano from the ground floor. They got it onto the verandah and in trying to manhandle it down the front steps it got away from them and rolled end for end, disintegrating as it went.

I went around the outside to my end of the building, which was burning hard. I was watching it when the Station Commander, W/C Croil, stopped for a few minutes next to me. Just then there was a loud bang in the area of my room, following shortly by another bang, in fact, 21 of them spaced equally apart. Croil barked, "What in hell is that?" I answered innocently, "I don't know, Sir." I knew very well for I had acquired one of the marine distress flares which we had carried aboard our East Coast flights and which, when fired by tearing off a tape, threw coloured flares into the air 21 times. The distress flare was in my suitcase in my room. I unobtrusively moved away from the Wing Commander long before the flare had delivered its 21st bang.

When it was all over most of us had lost all our gear. In my case this included log books, civil flying licences, civilian clothes, military clothes—in fact everything except the uniform I was wearing, plus a walking stick, which all officers carried at that time. In addition, our trunk and its contents were lost.

We were moved to a building called the "60-day hut," which became our temporary Mess. The burned building had been called the "28-day hut." We were issued with a suit of pajamas, all very large in size, a toothbrush and a few other necessities. One way or another, however, we managed to carry on. We all filled out claims for our lost gear and submitted them to higher authority. From then on, however, we lived in the one RCAF uniform we possessed. We could not change into civilian clothes in the evenings because we had none.

The course continued with ground school classes, blind flying, navigation flights and night flying. The blind or instrument flying was the first attempt at such a thing in Canada. The system involved flat rather than banked turns. Instead of banking the aircraft in the direction of the turn, as is now normal, we were instructed that when flying blind we must skid the aircraft around. This is most uncomfortable and unsafe. Our skids were timed to give us somewhere near the degree of turn we required. After completing the required time of skidding turns we then held our course straight and level to allow the magnetic compass to stop spinning and to settle, showing the new heading (there were few directional gyros in use then). When the compass settled we usually found that another small skid or two was required to put us onto the correct heading. The pupil's cockpit was covered by a folding canvas hood while the instructor sat in the other cockpit in the open.

Some of our blind flying was done in Puss Moths, small cabin aircraft with a pilot's seat in front and two passenger seats behind. The blind flying Puss Moths had their rear side windows blacked out and a folding instrument panel came down from the roof. This enclosed the pupil in the back and also gave him his own set of flying instruments.

We usually used the Gipsy Moth for our night flying, wearing helmets and goggles, heavy Sidcot flying suits, lined gauntlets and lined flying boots. Over everything else we buckled our parachutes. In spite of the heavy clothing we were stiff and cold when we climbed out of our open cockpits after a spell of night flying on a cold winter's night.

The electric power for the navigation and landing lights on the aircraft was supplied by a small wind-driven generator mounted on the forward edge of the upper wing. The generator had a miniature wooden propeller which was driven by the slipstream. The engines were not equiped with generators. In addition, a storage battery or two were stowed in the front cockpit. The night flying aircraft gave out an eerie scream in the cold night air due to the wind-driven generator, noticeable even to the people on the ground.

At this time the Station CO, W/C Croil, spoke to me about renting his house for a short time. He wished to move his family into Barrie to be closer to school, and to avoid having to use the often snowed-in roads. He knew that my wife was in Barrie and that it was difficult for me to get in except on weekends. We arranged that he would move his family into town and I would move Lin to Borden and take over the house, which, of course, was completely furnished. He retained one bedroom for his own use, for nights when snow conditions kept him from driving into town. This was a wonderful arrangement for Lin and me. Many pleasant evenings when I was not night flying were spent sitting around the fireplace with some of our friends from the Mess, enjoying a quiet noggin.

The course was a very good one. We were all pleased with it, particularly the blind flying, in spite of the flat turns. We realized that blind flying or, as it was called later, instrument flying was the coming thing.

Everything continued to go well until near the end of February or early March when the Federal Government, under the leadership of Prime Minister R.B. Bennett, decided to make some drastic cuts in spending. The RCAF was hit hard and was, in fact, cut almost in half. Our course was stopped completely at that point. The Reservists, such as myself, and the Non-Permanent section (a good percentage of the Force) were all to become inactive. Flying almost came to a halt. Camp Borden was paralyzed with shock. The members of our course were to be returned to civil life as promptly as possible and this at a time when the economic situation in civilian life was at its lowest ebb in the Great Depression.

Orders came out stating that members of our course could, if they wished, remain on duty for another month, minus any flying, to act as orderly officers and to paste amendments in various RCAF manuals. I was one of the group who decided to stay as it gave me more of a chance to get myself aligned with a new flying job.

My flying time acquired on the course was about average—48 hours and 30 minutes, which covered all phases of our flying—blind, navigation and night, both dual and solo. My closest associates on course were P/O George P. Silke of Vancouver and Sgt. Pilot Frank I. Young of Hamilton and Toronto. All three of us were having a tough time keeping ourselves fully employed in commercial aviation so we spent many evenings together discussing our future.

I wrote a letter to my old associates in Sydney proposing forming a new company (Maritime and Newfoundland Airways had ceased to exist). I suggested that we overhaul two of the Fokker Universals and fly them out to Edmonton and operate north into the bush from there. Mining was becoming important in the Northwest Territories as a result of some good finds by prospectors, so I thought it might be a good area to operate a couple of aircraft. I received a reply from Sydney agreeing with my idea and asking me to come down when I could and bring another pilot along. George Silke said he would like to go but Frank Young decided to stay around the Ontario-Quebec area where he had good connections.

The remaining members of our course all had financial worries of some degree or another. We were all deeply concerned that our claims for the loss of our clothing and equipment resulting from the Officers' Mess fire had not been paid. Authority had not been granted from Ottawa to make the payments and as a result we were all forced to buy civilian clothing with what little money we had. A meeting was held at which it was decided that some sort of pressure should be exerted on Ottawa to hasten the outstanding payments. Since George Silke and I were going to Sydney, it was decided that we were to be the representatives of the group and were to make a stop-over in Ottawa from our train journey in order to do something about the claims. We agreed to do what we could, although we were completely in the dark as to how to go about it.

Early in April, George Silke and I took leave of our RCAF instructors and, with Lin, departed from Camp Borden by train to Toronto, George and I wearing our only suit of civilian clothes. We could not afford to replace our lost suitcases, so we were carrying large brown paper parcels in which we had rolled our RCAF uniforms, spare shirts, socks and toilet articles. In our right hands we carried our RCAF officer's walking sticks, as we had not yet decided how to dispose of them.

It appeared to me that our new venture was going to be very dicey indeed, financially and otherwise, so Lin and I decided that she should go home to Medicine Hat for a while, until we could join forces later in Edmonton. George and I saw her off on her train, then we climbed aboard ours—next stop Ottawa.

5

Salvaging Kit and Careers

On arrival in Ottawa we registered at a small hotel near the railroad station, then started planning how to go about getting Parliament, which was in session to approve our fire loss payments. After due consideration we decided that the first step was to proceed to Rockcliffe Air Station, on the outskirts of the city and look into the matter from there.

We took a cab to Rockcliffe and introduced ourselves to the Adjutant, who took us over to the Mess to meet some of the officers and to have a drink. While enjoying our drinks the Commanding Officer, W/C Lloyd Breadner, came into the Mess. We were introduced and from that day on became good friends. During the Second World War he approved many of my proposals which were necessary to form an RCAF Air Transport Command, both in Canada and overseas. Breadner later became the Chief of the Air Staff (1940-43) and also Air Officer Commanding in Chief, Overseas (1944-45), with the rank of Air Marshal. He retired after the war with the rank of Air Chief Marshal, Canada's first air officer to achieve that rank.

During our first conversation with the Wing Commander we explained what we were trying to do in Ottawa. He was sympathetic, feeling strongly that the fire payments should have been made before we left Camp Borden, so he decided to help us. Since the House of Commons was in session, he accompanied George and me there for the night session. We hoped to be able to speak to a member or two to solicit their aid in raising the matter on the floor of the House. Breadner knew a couple of MPs very well and George knew a member from British Columbia, but I did not know a single one.

The Wing Commander changed into civvies, we had dinner together and went to the House of Commons. We found George Silke's friend in his office and he took us to a larger room where there were a number of MPs. We were all introduced, supplied with a drink and given the opportunity to explain our problem. The members were wholeheartedly in favour of something being done. We had a few more drinks and about this time the bells rang to summon the MPs into the Commons. They arranged for us to sit in the gallery, escorted by a cabinet minister.

George, the Wing Commander and I were by now feeling the effects of our drinks but were quietly behaving ourselves. There was much discussion on the floor about the program for the evening and one or two of our new friends jumped to their feet, trying to get a discussion started about our Camp Borden fire losses. This caused quite a commotion and in the end the Prime Minister ruled that since it was not on the agenda it should not be discussed. With that, George and I jumped to our feet and shouted, "Down with Bennett." In no time at all we were hustled to the front door. The minister and the Wing Commander joined us outside.

We were a bit dismayed at what we had done but there was nothing we could do about it now. In any case, strangely enough, each member of our course received a cheque for payment in full of fire losses just one week later, so apparently our disturbance in the House paid off.

At this point I described a pleasant wine I had run across on St. Pierre, so without further ado we all proceeded to Hull to find one like it. We sampled many drinks but I cannot remember whether we were successful in our quest for the special wine. We were dropped off at our hotel very much under the influence of alcohol, but not until it was decided that we should get together and work on the wine situation further the following night.

George and I woke up with the worst hangovers imaginable and decided that, fire losses or no fire losses, we could not face another evening like the last one and that the only thing to do was to leave town as soon as possible. We found, or thought we did, that a train was leaving for the east shortly.

We walked shakily into the station, our throbbing heads making sound thinking impossible. We each carried our brown paper parcels containing our gear, and our walking sticks. The sticks were becoming a real nuisance by now but we did not wish to throw them away. We arrived on the platform to see a train in the act of pulling out and fuzzily assumed that this must be ours. I said, "Let's go, George," and we started to run. Being taller than George, I managed to reach the rear platform of the train and scrambled on. George was still trying to make it but was losing ground. Suddenly he stopped and waved frantically for me to get off. I jumped but the train was going faster than I thought. I hit the platform, my walking stick got tangled in my legs and I spread-eagled myself on the station platform. My brown paper parcel burst open, scattering things in all directions. People on the platform were laughing their heads off.

George told me that while running he had seen something on the rear of the carriage which convinced him that that was not our train for the east, so he had stopped and waved to me to get off. We gathered up my belongings and re-tied them. On making inquiries we found that our train was leaving shortly, so we made damned sure this time that we were on the right one.

After a journey undisturbed by further difficulties we arrived in Sydney. We went directly to the home of Mrs. McDonald, where Lin and I had lived, and asked if we could rent accommodation. She was pleased to rent us a large room with two beds, for a very reasonable sum. I contacted Dr. Rice, a well known and respected medical practitioner, now heading the small group that owned the Fokkers. Dr. Rice had been the chairman of Maritime and Newfoundland Airways and was now directly in control of everything, since our manager, Bobby McGowan, had partly retired from the scene.

A meeting was arranged for the following day where Dr. Rice introduced George and me to a small group of Sydney businessmen interested in forming a new company. After much discussion it was decided to form Explorers Air Transport Ltd., headed by Dr. Rice, with myself as chief pilot. The aim was to overhaul two of the Fokkers thoroughly, load them up with spare parts, ground handling gear, tools and so on, and fly them to Northern Alberta and the Northwest Territories, there to establish a suitable base, and get into the business of flying mining engineers, prospectors and others. Recent discoveries of silver and pitchblende by the Labines had created much interest in the North, in spite of the crushing shortage of money created by the Depression.

The two best Fokkers, G-CAHE and G-CAHJ, both required considerable repairs to their plywood wings, some new sections to be welded into the tubing of the metal fuselages, new linen fuselage covers, engine overhauls and repairs to the old Hamilton floats. In addition, work was necessary to put in good shape a complete ski undercarriage for each machine for winter flying.

A good woodworker was hired to repair the wings, but George and I had to do all the rest of the overhaul program. Our Air Engineer's Certificates, under which the operation was done, were only good for the "Inspection of Engines and Aircraft before flight," and were not good enough for a major overhaul. This caused complications later on.

Actual work began around May 1, 1932. George and I would drive back and forth between Sydney and North Sydney where our hangar was located, in a borrowed car or by bus.

Sufficient money was provided by the company to pay the wages of the woodworker and to buy parts and materials as required, but there was not enough for any salaries for George and myself. We had to manage, using our small savings acquired from our pay at Camp Borden. There were no other aviation opportunities available to us, so to stay in our chosen field it was necessary to carry on just as we were doing.

We had to supervise the repairs of the wings as the carpenter was not an air engineer; and also learn how to cut out pieces of corroded tubing from the fuselages, insert metal sleeves of a smaller diameter at the joints, then weld in the new pieces. We both learned to do fairly good welds which must have been OK because the fuselages later survived much rough air without coming apart. Ensuring that the fuselages were true after welding was also essential. They were recovered with new linen, which was sewn, fitted, treated with aircraft dope, then painted. We also did our own wing and fuselage lettering.

The engines were torn down and cleaned, valves were reground, magnetos were checked and new contacts fitted, carburetors stripped and cleaned, and then everything was reassembled and adjusted. The leaky old floats required new rivets in many places, plus cleaning and painting. We worked from first thing in the morning until quite late at night because the sooner the job was done the sooner we could take off for northwestern Canada. We were a very tired pair.

One evening Dr. Rice suggested that we take the evening off and join him at his home for a few quiet rums. Rather late at night we left for our own quarters feeling very mellow indeed, with a fair supply of rum under our belts. At this time in Sydney, conditions were pretty rough, with very little money and many unemployed. Consequently the streets, except in the brightly lit centre of town, became dangerous late at night. There were hold-ups, beatings and even some killings taking place.

When we arrived at our quarters, away from the bright lighting, George went straight to bed while I decided to walk a few more blocks to work off some of the rum before turning in. I lit a cigar and started on my way. Shortly I heard a rumpus going on across the street and discovered three men scuffling with a woman. The men immediately turned and came at me. Applying every trick of boxing I had learned from Fergie Fraser in Medicine Hat, I retreated down the shadowy street, in order to keep them from getting behind me. I was taking a heavy beating, particularly from my attackers' kicks. A face appeared as we passed near a lamp post, so I hit it as hard as I could and it disappeared into the gloom. Someone from one of the houses nearby had apparently phoned for the police because a cruiser screamed around the corner. The three men vanished into the darkness and left me standing there, quite unable to move any further.

The policemen drove me right back to the place I had left a short time before, the home of a surprised Dr. Rice. He examined and patched me up. I was cut and bruised in a number of places but the worst damage was a broken bone in my left foot, which never did set quite right and which bothered me for a long time afterwards.

After another drink Dr. Rice gave me a set of crutches and the police drove me home. Needless to say, George received quite a shock when I awakened him to tell him what had happened. After a few days of rest, I was able to return to work, but found my old walking stick necessary to help take the weight from my sore foot. The police informed me later that they had arrested the men when one of them went to a doctor for treatment of a broken jaw. I felt somewhat better about the whole thing.

Once or twice during our overhaul a Civil Aviation Division aircraft arrived with engineering inspectors to review our progress. The Division had learned of the overhaul and was concerned about our lack of an Engineer's Overhaul Licence. However, they did not wish to stop the operation and add to the number of small businesses which were closing their doors due to lack of funds. They had watched our work so carefully that we were permitted to sign the aircraft out for a test flight.

By the evening of June 12, we had completed the overhaul of G-CAHJ, which I decided would be my aircraft for the western flight. The next morning, we rolled the glistening Fokker seaplane down the slipway on its dolly and slid it into the water, where it floated clear. We refuelled it and checked the engine oil. Bobby McGowan, our old manager and now a member of the new board, wanted to go with me on the test flight. Bobby loved airplanes and flying and rarely missed an opportunity to go along. I told him to get aboard and then started the engine.

Once again I was standing up on the seat in the open cockpit and fitting the starting crank into the dog, which projected through the top of the cowling ahead of the cockpit windshield. After winding the crank until the starter flywheel was turning fast enough, I snatched out the crank, dropped back into my seat, switched on the ignition and pulled the starter control. The J-4B engine turned over a few times, then we moved out into the harbour. In 1932 there were no float rudders to steer the aircraft on the water, and I had to manoeuvre with my tail rudder.

I taxied down the harbour, allowing my engine to warm up, and observing that oil pressure and engine temperature indications were satisfactory. I tried the flying controls, noting that the rudder and elevators were free and that the ailerons moved in the correct direction when I moved the control stick. I turned into wind, slowly opened the throttle wide and quickly checked engine revolutions on both magnetos by using the ignition switch. Everything worked properly so I let the seaplane come up onto its step, plane along the surface and become airborne.

It was a wonderful feeling to be airborne again after all our hard work and troubles of one kind or another. I found in the climb, however, that the controls did not centre properly and that I had to fly with the control stick off centre, but this we would readjust when I landed. Other than that the aircraft flew beautifully. After 15 minutes I descended and landed with a bit of a bounce. It was over six or seven months since I had flown on floats but, all considered, I was satisfied.

The next day, after making our adjustments, I flew again for 30 minutes and all was well again. On the 15th, an hour's flight was done; however, engine trouble developed and a forced landing was made for repairs.

George's machine, G-CAHE, was now finished and wheeled out for flight. He delightedly climbed into his cockpit for the test flight. Bobby McGowan, as usual, clambered aboard and took his seat in the cabin. George did an excellent test flight with a reasonably good landing at the end. His machine was working well.

At this point the Civil Aviation inspectors came down again and were pleased that the test flights had been successfully carried out. They decided to overlook our lack of overhaul qualifications and let us carry on normally. They were a helpful group of people.

Flights and adjustments were carried out each day as well as finishing the overhaul of the ski equipment, which was to be shipped by rail to Edmonton. But on the night of June 19, my 26th birthday, trouble hit us again.

I was working at a bench in the hangar, grinding some metal from a fitting with an electric grinding wheel. Carelessly, I had not put goggles over my eyes and before I knew what had happened, I had a number of pieces of steel imbedded in my left eye and was in a fair amount of pain.

We closed the hangar and I immediately went to see Dr. Rice. He removed some of the pieces of steel but found one piece deeply imbedded near the iris. This, he said, required the services of a specialist. There was an eye specialist associated with the steel mills in Sydney who was a friend of Dr. Rice, but he could not see me until the following morning. We were all upset as we were almost ready to start on our western flight. In addition, Dr. Rice was concerned that I might lose the sight of my eye.

The next morning Dr. Rice, George and I went to the office of the specialist. An operation was conducted and my left eye was to remain bandaged for some time. The specialist assured me that all the steel pieces were out, but whether the eye was ruined or not we would not know until the bandage was removed later. I was in severe pain, so was given a couple of large drinks of neat rum which had little effect to begin with, but which gradually made life a bit more pleasant.

We returned to Dr. Rice's and discussed the matter of my eye and our imminent departure for the West. I knew that it was quite illegal, as far as Federal regulations were concerned, to fly commercially with the sight of only one eye, but since my friends in the Civil Aviation Division were not aware of my mishap, I recommended that we leave as planned. After much discussion and some opposition from Dr. Rice, we agreed to leave early the next morning, June 21, 1932. I hoped I would be able to see well enough with one eye for my coming take-offs and landings.

6

Edmonton or Bust

We had loaded the cabins of our aircraft with as many spare parts, tools and other gear as possible, in addition to an iron anchor weighing about 40 pounds and a large coil of anchor rope for each aircraft. These anchors were awkward to handle. When the anchor was needed the pilot, when operating alone, had to climb from the open cockpit and down the side of the aircraft on metal steps, duck under the wing struts, climb into the cargo cabin, carry the heavy anchor and rope down to the float, duck under the wing struts again, go to the nose of the float and tie on the anchor, then throw it overboard. This procedure was the cause of a touchy but humorous episode later in our travels.

We took off from Sydney harbour early on the morning of June 21 into low cloud and rain, but we did not expect too much trouble. Our first stop was planned for Saint John. I found that I could see sufficiently well with my one eye that I had no trouble on the take-off, and later that I had no real difficulty in landing unless conditions were abnormal. The weather gradually worsened as we flew along, until we were forced to climb above the low cloud deck. I had found that my aircraft was about 10 mph faster than George's, which meant that I had to either throttle back to stay with George or, rather than have my engine operate below normal revolutions, fly ahead and then circle until he caught up.

When our estimated time of arrival was up, indicating we were near Saint John, we could not find a break in the clouds below so turned back to the edge of the cloud deck. We landed on the Petitcodiac River near the Bay of Fundy as the huge tide was coming in. We dropped our anchors but the bottom of the river must have been mostly mud and the power of the tidal bore made us drift fast with the current. The anchors did not even slow us down so we pulled them up and took off again, finally landing at Port Elgin, New Brunswick.

By this time our fuel was low and, as none was available in Port Elgin, we partially filled our tanks with automobile gasoline, hoping that this fuel mixed with what we had left would keep the engines running. A telephone call to Saint John revealed that breaks were now showing in the clouds and that we should arrive as soon as possible before it started to close in again.

Our take-off was successful and we were again airborne for Saint John. The fuel we had taken on at Port Elgin was only sufficient to take us to Saint John but not to return again. Our engines, to our relief, seemed to be operating well with the mixed fuels, but strong headwinds forced us to climb on top of the cloud again. We arrived where Saint John was supposed to be but found that the reported cloud breaks had disappeared and it was now solidly undercast. We had no radio communication, of course, in those days, so we waved to each other from our cockpits and began to circle, looking for a break. Meantime our fuel was so low it had ceased to show in our fuel gauges. I wondered how I would manage during the coming emergency with the sight of only one eye.

Suddenly my engine stopped and I waved to George and headed down for the cloud. I thanked heaven that we had just finished an RCAF course at Camp Borden on blind flying, even though at that time we used the flat turn system. I went into the cloud, keeping my air speed up and hoping there were no hills in the area. After what seemed like a long time I suddenly broke through and was over water at low altitude. I quickly flattened out and landed and could then see Saint John not too far away. My one-eyed flying had worked out all right.

After removing my helmet I could hear George still circling above. Suddenly his engine stopped and in a minute or so he popped out of the cloud and landed, by coincidence not too far from me. Boats darted out and towed us into the dock at Millageville, near Saint John. We were a very relieved pair when we stepped ashore. Flying time for the day was 6 hours and 30 minutes per aircraft.

June 22 was again a day of low ceilings and rain so we spent it refuelling and checking over our aircraft for departure on the following day. On June 23 the weather looked better so an early take-off was made in hopes of getting as far as Montreal before nightfall. This was not to be, for while following the St. John River and after passing over Bath, New Brunswick, we met rain and hail and were forced to turn back to Bath. As senior pilot it was my job to land first. There seemed to be little or no wind (the calm before the storm) as I made what I thought to be a good approach and landing, and taxied to shore, where people were clustered on the bank watching the operation. They helped me tie the aircraft up.

George, meanwhile, was making his approach from the opposite direction, no problem due to the calm conditions. Suddenly I saw that he was heading towards a high cable stretched across the river. Just in time he pulled up sharply and missed it, continued on and landed. After he pulled into shore I questioned his action in not using the approach I had used, and which I considered safer. At this point a local resident spoke up and told me that I had just missed one or two cables by a matter of a few feet while doing my final approach! I could not believe it, but the man drove us along the river road in his car and showed us the cables. There was no question but that I had nearly hit them, probably due to my bandaged eye. George and I looked at each other, laughed and decided to drop the whole matter.

On returning to the aircraft we noticed that the rear float struts on one side of George's aircraft were badly out of line. These struts were fastened by a clamp onto the rear wing struts and this clamp had slipped. The rest of the day was with repairs which required blocking up one side of the aircraft along the shore, ably assisted by many of the local townspeople. Flying time for the day had been two hours per aircraft.

The next day, with no improvement in the weather, we passed the time touring the town with some of the people we had met. Always optimists, we gauged the weather satisfactory on the 25th and took off, again hoping to get as far as Montreal. Once more we were frustrated. The ceiling gradually dropped, heavy rain developed and the visibility finally became so poor, even though we were flying close to the tree tops, that a forced landing became necessary. A suitable lake suddenly appeared so we hurriedly alighted. The shoreline appeared a bit rough and we had to anchor the machines out on the lake, which necessitated going through our complicated anchoring procedure. When this was completed a boat with two men in it appeared and took us ashore. We had been airborne only 45 minutes that day.

We had assumed that we were still in Canada, but on getting ashore we were met by a uniformed United States Customs or Immigration official who pointed out that we were just inside the US border, at a town called St. Agathe, Maine. This caused some consternation as we had not, of course, given official notice of our pending arrival, nor had we been given permission to enter the USA. We could not take off again due to the weather. Finally it was accepted that this was an emergency landing and that we would depart as soon as weather permitted. We spent a very pleasant eventing with the US official, his friends and a few bottles of beer.

Very early the next morning, we pulled up our anchors, started our engines and took off. The weather was fair and everything seemed to be going well. We reached the St. Lawrence River near Rivière du Loup and headed for Montreal. I noticed an occasional fluctuation in my engine oil pressure, so from there on watched it closely, but after passing Quebec City the pressure dropped quickly. I waved to George that I was going down, so he swung in behind and followed me. We landed on the St. Lawrence and beached at Port Neuf. Flying time for the day was 2 hours and 45 minutes per aircraft.

Considerable work was necessary to rectify the oil trouble and it was not until the morning of the 28th that we continued on. We landed on the river near Longueuil, close to the Fairchild factory, and tied up after 1 hour and 20 minutes flying time.

We stayed in Montreal until the morning of June 30 because of poor weather. Whilst in Montreal I removed the bandage from my eye and found that I could see, not too clearly at first, but gradually better as time went on. We celebrated the event in the proper manner by hoisting a few drinks. For two or three years afterwards I could feel a slight ridge of tissue or something on my left eyeball but it finally disappeared.

On the morning of the 30th the flight was resumed for Trout Mills, near North Bay, Ontario, and other than bucking fairly strong head winds the trip was a pleasant one. We landed and tied up after 4 hours and 30 minutes in the air. The afternoon was spent in procuring aviation gasoline, refuelling and making ready for the next hop.

We were running very short of funds and, although we had wired Sydney for more money, it had not yet caught up with us. Our expenses were high as we had to pay in cash not only our travelling expenses, but also the aviation gasoline and oil used by the aircraft, in addition to docking charges and repairs. As the next day was July 1, and a general holiday, we decided to do some barnstorming to raise money. We flew to Lake Nipissing near Callander and carried a few passengers but did not manage to improve our finances appreciably.

On the morning of July 2, we took off from Callander for Sault Ste. Marie, bucking strong head winds, and on arrival over the Sault we found water conditions to be poor due to the winds. I signalled George that I would go down first, finding it very rough on landing but safe enough. Wallowing in the waves I held my position by use of the engine while waiting for George to land, and also for someone to appear on shore to give us a hand beaching the aircraft.

George landed successfully nearby but apparently decided that the beach was not satisfactory and that he would anchor off shore. I watched him switch off his engine and then realized what he intended to do. I waved from my cockpit to get him to start his engine again but he did not see me. What happened then was serious at the time but comical in retrospect.

George scrambled hurriedly down the side of his aircraft to the deck of the float, as he was drifting fast and rolling heavily. He ducked under the wing struts and disappeared into the cabin, backing out again carrying the heavy iron anchor and the coil of rope. Again he had to swing under the wing struts of the rolling aircraft, this time with his heavy load, which he managed successfully. Finally he staggered to the nose of the float and, noticing how fast he was drifting, quickly threw the anchor overboard, having forgotten to tie one end of the rope to the float!

Suddenly he realized what he had done, but there was no time to tie the rope now. He rushed back, with the end of the rope to the wing struts and, holding the rope with one hand, threw the other arm around the wing strut. The rope quickly became taut. The aircraft turned almost sideways, jerking George off the float and, still holding onto the rope, he was pulled right up to where the wing strut joined the wing, and there he hung over the water. He realized that there was only one thing to do and that was to let go of the anchor rope. With that, he slid back down to the float and the aircraft weather-cocked into the wind. In a moment he drifted tail first onto what turned out to be the smoothest part of the beach.

At this time pilot George Phillips of the Ontario Provincial Air Service arrived at the beach with an assistant. He helped get George's aircraft tied down and then signalled to me to drift in alongside. We finally had them both secured. Phillips then kindly got a boat and we went out and located George's anchor and rope and brought it back in. He then arranged for some aviation fuel for us and took us to a hotel. (George Phillips won the McKee Trans-Canada Trophy in 1931.) Flying time per aircraft was 3 hours and 30 minutes that day.

We now realized that we were in trouble as we were completely out of money. We still owed George Phillips money for refuelling our aircraft and were obviously facing hotel expenses. This situation existed for six days which, of course, turned out to be six good flying days, but we were unable to leave as we could not pay our bills. We sent frantic collect wires to Sydney for money. In the meantime, we were forced to eat comparatively expensive meals in our hotel because they were being charged on the bill. We could not go out and find cheaper meals as we had no cash. Finally, on July 7, some money arrived. We paid our bills with not too much to spare and prepared to leave the next morning.

On the morning of the 8th, although head winds were strong, we decided to leave. We said goodbye to George Phillips, who had been so very helpful, and took off using the direct and shortest course across the middle of Lake Superior in order to save money.

All went well until we were about three quarters of the way across the lake when I noticed that George's aircraft would frequently lose altitude almost down to the heavily rolling water, and then would slowly climb again. I was baffled by this until we compared notes later, when he told me that his engine was cutting out and then picking up again just before he had to consider ditching.

About 25 to 30 miles from Fort William, the weather closed in and for the first time we lost each other. I circled around very low but could not find George, so I continued on to Fort William where I landed and tied up to a dock, very concerned about him. A dockhand said he had not seen or heard George's aircraft. I then asked about getting some aviation fuel as my tanks were very low, but was told it would take about an hour. I felt that I could not wait that long as George could be down on the water and in trouble, so I told the man what had happened and asked him to despatch a launch if he could.

I took off again, with little fuel in my tanks, flew to the area where we had been separated, but could not find any sign of George and his aircraft, so I headed back for Fort William. Just as it was coming into sight my engine died from lack of fuel. I made a rough forced landing, then went back into the cabin, via the wing struts, and took out two of the marine distress flares we carried (the same type that was in my baggage at Camp Borden during the Officers' Mess fire). I fired both, attracting the launch that towed me into Fort William. I was very upset as I was sure by then that George had crashed into the lake. I mentioned this to the boat crew and they said that just as they were putting out to help me they had heard that an aircraft similar to mine had landed near Port Arthur, a little further along the lake shore.

After tieing up my aircraft I took a taxi to Port Arthur and, sure enough, there was George's aircraft tied to a dock. I found him trying to get some aviation fuel to go out to find me! He thought he had landed in Fort William and since I had not shown up he was sure that my aircraft had gone down. We were both so relieved we consumed quite a few beers to celebrate our reunion, in spite of our financial woes. Flying time on this flight was 5 hours and 15 minutes, plus the extra time involved in searching for George, which was not recorded.

The next morning, July 9, after refuelling both aircraft, we were again out of money. Our next planned stop was at the RCAF seaplane base at Lac du Bonnet, Manitoba. We took off from our separate locations, joined up in the air and headed west.

Headwinds increased in strength as we flew and finally we ran into heavy rain. As visibility was gradually getting worse, we landed at Kenora, Ontario, and tied up. One of the first people to come to the dock was "Shorty" Keith, whom George and I knew as F/O Alvin Keith, RCAF, and who was at Camp Borden with us and had also been transferred to the RCAF Reserve of Officers. Shorty was flying a special aircraft for a large food processing firm and had flown the top executive of the firm to Kenora. After explaining to Shorty that we were out of cash he arranged that we stay with him in quarters rented by his company, which was a big help. Flying time for the day was 3 hours and 20 minutes.

July 10 was a good flying day. We said good-bye to Shorty Keith (he was killed later in an air crash) and took off for Lac du Bonnet. There we tied up at the RCAF base and two RCAF friends drove us into Winnipeg, as we had decided to try to get a short airlifting job for our aircraft in order to raise some money. Flying time for the day was 1 hour and 10 minutes.

The next six days were spent in Winnipeg. We had wired Sydney for more money and were also making the rounds of the existing aviation companies trying to get a small flying contract. The Depression was at its peak in that area and there was simply no work for us. The day was finally saved, however, when some more money arrived from Sydney. We paid our bills and planned our next flight, north to The Pas in order to follow the most suitable water route to Edmonton. This was necessary as we had no desire, of course, to cross the prairies on float-equipped aircraft.

We took off from Lac du Bonnet on July 16 and headed north. The flight was uneventful and we landed and tied up at The Pas after 4 hours and 15 minutes. The next day, we took off for the RCAF base at Ladder Lake, Saskatchewan, where we arrived after 3 hours and 15 minutes. After taking on fuel we decided to push on but 45 minutes later we ran into electrical storms which necessitated landing on a reed-encircled lake at Spiritwood, Saskatchewan. Kindly people showed up and helped us secure our aircraft for the night.

July 18 was a clear day. We took off and headed west, and after 3 hours and 30 minutes we put our aircraft down on Cooking Lake, just east of Edmonton. This was to be the southern base for our operations into the North.

The first part of our job, to fly the aircraft from Sydney to Edmonton on floats was completed. We were 27 days en route and each aircraft had flown 43 hours and 20 minutes. Next we had to organize our new western branch of the company, arrange flying contracts, establish expensive northern fuel caches, and get into operation—all this with little or no money and in the middle of the Depression.

7

Survival Training

After carefully securing our aircraft tail first on the beach at Cooking Lake, George and I went into Edmonton and took two rooms at a local hotel, which was much used by mining men going north, and where we expected to arrange some flying business. Lin arrived from Medicine Hat, where she had been staying with her parents. It was great to see her and have her support again. She was always loyal and cheerful, regardless of the circumstances, and this was the spirit much needed at this time.

There did not seem to be much prospect of business at this stage so we decided that George should fly his aircraft to Prince George, British Columbia, where he had previously done commercial flying for mining operations. He thought he could generate enough business to support one aircraft in that area. I decided to try scrounging enough business to support my aircraft in the area from Edmonton north.

After George took off for Prince George we moved out of the hotel for reasons of economy and took a small summer cabin at Cooking Lake. The cabin was one of many in the area owned by an old gentleman named Charlie Upright. We moved our few belongings out there where I was now able to keep an eye on my aircraft but I had to get in and out of Edmonton in order to pick up business.

Money was again short, so I tried my old prairie barnstorming pattern, this time on floats from the surface of lakes. I concentrated largely on Seba Beach at Lake Wabamun, west of Edmonton, which was quite a summer resort and provided some business to help out. Lin and I slept in the aircraft on coils of rope and the canvas engine cover during these forays. On one occasion we were accompanied by a mining engineer named Chuck McLeod, prior to my flying him, his men and equipment to Fort McMurray, Alberta, on their way north.

In order to operate aircraft north from Fort McMurray to the Arctic regions, it was necessary for each company to have its own aviation gas and oil supply at each major stopping point. Fuel was very costly by the time it was transported into the north and caches were arranged at the various bases. Consequently, one company would not permit another to use its fuel, an ideal means of eliminating competition. It was necessary, therefore, that I have suitable fuel caches, otherwise, I could not accept any business.

Over the signatures of company people in the east, as well as my signature, the oil company agreed to supply aviation gasoline and engine oil. I also secured the help of Jim Darwish, a businessman who owned a fur trading post at Fort Rae, Northwest Territories, and who also transported goods into the area north of Fort Smith. Jim agreed to ship my gasoline on his boat, setting up caches at Fort Smith, Fort Resolution and Fort Rae, but at this season there was no way of getting the supplies further north to Great Bear Lake. Unfortunately, on his first try with the gas his boat foundered in a storm on Great Slave Lake, with the total loss of my Fort Rae supplies. He secured another boat and got more fuel into Fort Rae just before the winter freeze-up.

It was impossible for me to take advantage of my northern fuel supplies as the caches would not be completely in place until freeze-up, so my earliest operations north of Fort Smith would have to be after freeze-up and on skis in place of the aircraft's floats.

I managed to generate a few trips from Cooking Lake to Fort McMurray carrying prospectors and their supplies and sometimes, with luck, finding a partial return load. Fort McMurray, a small settlement near the junction of the Athabasca and Clearwater rivers, was used by the mining and aviation companies as a jumping-off place for the north. It was just three miles from the town of Waterways, which was at the end of the railroad from Edmonton. At that time, a train nicknamed the "Muskeg Limited" operated once weekly between Edmonton and Waterways.

About this time, George Silke flew in from Prince George for a discussion as to our future operations, bringing with him his wife, Gertrude. We were certainly pleased to see them both, as they were and are two very fine people. They took the summer cabin next door to ours for a few days. Business was not very good in the Prince George area and George was concerned about the future.

From August 26 to September 22 I managed to fly four return trips to Fort McMurray and one to Lac La Biche and return. On the return flight from Lac La Biche on August 27, I was flying an empty aircraft and steering across fairly dry country when my engine started cutting in and out. The only possibility of a float landing on water was on Smoky Lake, directly on course ahead of me. I lost altitude pretty fast and finally my engine stopped completely. There was silence except for the windmilling propeller. I managed to stall over the edge of the trees and down onto the lake surface, but my aircraft seemed to be held back as though by brakes as I touched the lake surface, almost flipping over nose first. It actually cocked its tail fairly high in the air and then flopped back again on its floats. I cracked my head on the instrument panel, which dazed me for a while.

I found, on recovering my wits, that where I had landed was very deep mud with only shallow water above it. Some people showed up on the shore and shouted to me to stay where I was. They brought some planks and finally a boat or two came over from the deeper water further out in the lake. Using the planks, I managed to get at the engine and rectify the trouble.

With the aid of people, planks and boats, and finally my engine, the aircraft was moved into deeper water. I washed some of the mud from it and myself, and found that everything seemed to be all right. I said goodbye and thanks to my friends and managed to take-off and return to Cooking Lake.

The next day, August 28, I flew a load to Fort McMurray, where I remained overnight, trying unsuccessfully to get a return load. Just then Walter Hill, owner of the local drugstore, came down and asked me if I would fly him and his wife, Gladys, out as soon as possible on an emergency basis. She was due to have a baby and the local doctor believed there were possible medical complications. He recommended an immediate flight to Edmonton where proper facilities were available.

The aircraft was ready to go so Mr. and Mrs. Hill were brought aboard and strapped into the metal seats in the little cabin. I took off, climbed out of the valley to 2000 feet and headed for Edmonton, staying within gliding distance of the Athabasca River, at least as far as the town of Athabasca Landing. From there on it was a very dry flight into Cooking Lake, due to lack of lakes and rivers in that part of the country. All went well for a while and finally we ran into lowering ceilings and intermittent rain. About this time Walter Hill opened the small peep door between the cabin and my cockpit. He tapped my arm, so I undid the strap of my flying helmet, raised one side of it and placed my ear as close as I could to the door. Walter shouted to me that Gladys was in pain and to get to Cooking Lake as fast as I could. I nodded and opened my throttle wide, increasing my air speed to its maximum, which was about 100 miles per hour.

I had to get closer and closer to the tree tops due to the weather, while keeping an eye on my old engine which was doing its utmost. Finally Cooking Lake showed up and I went in, landed as quickly as possible and taxied to the little dock near our summer cabin. On hearing the engine, Lin, Gertrude and George Silke all came to the dock and helped me tie up. Walter Hill said that Gladys was going to have the baby any minute so we carried her into our cabin and put her on the bed. Walter Hill, Lin and Gertrude took over, while George and I were told to go outside and call for medical help from Edmonton.

Two doctors and two ambulances arrived, but by then everything was done. The baby had been born ten minutes after we had landed but Walter, Lin and Gertrude had done their jobs well and no complications developed. They had made the baby's first clothes from pillow cases. On arrival the doctors said that everything was medically satisfactory so Gladys, the baby, Walter and the doctors all climbed into one ambulance and drove into Edmonton. Gladys and Walter are still in Fort McMurray. I do not see them very often but when we do get together we all enjoy each other's company to the full. This includes the baby, Kenneth Rowland Leigh Hill, who is now a family man in his own right. His older brother, David, another fine boy, was shot down and killed over Europe while on a bombing mission to Germany in World War II.

After about a week's visit, George and Gertrude took off for Prince George. We had decided that if we had to go broke that we would at least carry on to the bitter end, George in British Columbia and I in Alberta, the locations in which each of us stood the best chance of survival.

During September I made two more round trips to Fort McMurray, one of which was to take Walter and Gladys Hill and their new son home again. Even the small amount of business available was falling off as freeze-up approached. In my one-man operation as manager, pilot, air engineer, loader and unloader of the aircraft, I was now going to be faced shortly with a one-man operation of changing over from floats to wheels or skis at a lake where no landing strip existed nearby. It would require an ice surface on the lake strong enough to hold my aircraft.

In October it was getting very cold in our little summer cabin, which was heated only by a small coal stove, and we would frequently awake in the morning, shivering with the cold. I decided that it was time to move out. Once more friends came to our rescue, this time in the form of Jim and Evelyn MacDonald and their two sons. Jim spent much of his time in the north working for the Ryan brothers at Fort Smith and their two boys had accompanied me on some of my local barnstorming flights as helpers. Knowing that we were short of money, the MacDonald's asked Lin and me to stay with them in Edmonton for a while. It was very kind of them and it was nice to be in a warm building again. We stayed with them until the beginning of November and were always thankful for the help they gave us at that time. We were able to pay for our share of the groceries but that was our limit.

Grant McConachie, who later became president of Canadian Pacific Airlines, was just getting started with his own company. He had two Standard Universals similar to mine, except that a metal top had been fitted over the normally open cockpits. At this stage, Grant had one on wheels, G-CAGD. He called to ask if I would fly a couple of trips for him in it while I waited for the freeze-up. I was pleased to help. Our finances were still at rock bottom, as I did not receive a salary from Explorer's Air Transport, only a percentage of revenue, which was barely enough to get by on.

The first flight was to fly a couple of passengers from Edmonton to a number of points in Alberta—Cereal, Empress, Medicine Hat, Calgary and return to Edmonton. This flight was carried out on October 27 and 28, 1932.

The next flight, in Fokker G-CASE, was to take a small party involved in the movie industry to the Peace River area. We flew from Edmonton to Fahler, Grande Prairie, then back to Fahler and then to Edmonton. Winter had set in so that I had to heat up the engine with a plumber's fire pot each morning of the four-day trip. There were no aerodromes, so farmers fields were used as landing areas.

On my return to Cooking Lake I got busy changing G-CAHJ over to wheels. I managed, with such equipment as empty gas drums and jacks, to get the floats off and the wheels on. On November 9 I took off from the frozen lake and landed at Edmonton airport, which was to be my winter base.

There was no business at all for my aircraft so I decided to fly with Lin to visit our families in Medicine Hat and Lethbridge. November 10 was a cold winter day but we crawled into our unheated aircraft and took off. Weather forced me to land in Calgary but I was able to take off later, landing in Lethbridge just before dark.

The period from November 11 to just before Christmas was spent working on the aircraft, including a top overhaul on the engine. In addition, I had a number of meetings with Ben Metcalfe, a retired Lethbridge businessman and an old friend of the family. After some persuasion I talked Ben into coming back to Edmonton with me as my traffic and freight manager. I had found that I missed getting business at times while away flying or doing engineering work on the aircraft. We decided that we would set up a small office in the Royal George Hotel, which in reality, would be Ben's room. He would then be in a position to spend full time digging up business while I could concentrate on the flying, air engineering, and loading and unloading the aircraft. We agreed to fly to Edmonton shortly after the New Year, 1933.

My aircraft overhaul work was finished and the aircraft test flown satisfactorily just before Christmas. My brother, Bert, an accountant with the Bank of Montreal at Castor, Alberta, about midway between Calgary and Edmonton and somewhat to the east, was planning to come to Lethbridge for the Christmas season, as were two other young banker friends of his from Gadsby and Delia, not far from Castor.

Feeling the need to do some flying, I suggested that the three young bankers split the costs of aviation gasoline required for the trip and that I would come north, pick them up and return them later to their banks after Christmas.

I flew north on December 24 over a snow-covered landscape, found a suitable landing place near Gadsby, picked up the first banker, then flew to Castor to get my brother. The snow was a bit deeper there and I had some trouble getting in and out of the farmer's field on my wheeled undercarriage. I picked Bert up and then we got in and out of Delia successfully. They were pretty thoroughly frozen by the time we landed at Lethbridge. We all had a pleasant Christmas and I managed to fly them all back to their posts on December 26. One of the young bankers was Lawrence Stickley, who joined the RAF before the Second World War and became a well decorated group captain at war's end.

I received word that I had been elected, as of December 1, 1932, a member of the Guild of Air Pilots and Air Navigators of the British Empire. It was an honour to belong to such a group. Later a certificate arrived, signed by the Marquis of Clydesdale as Deputy Master and by Lawrence Wingfield as Clerk. Both were well known in early British aviation circles.

On January 5, we were ready to start for Edmonton and get back into northern flying. Lin and Ben Metcalfe got into the cabin and we took off. We got as far as Nobleford, just north of Lethbridge, when I ran into bad weather and had to return. The weather did not improve until January 8 when we tried again. Once more we got as far as Nobleford, and this time we encountered trouble with the aircraft and had to land in a farmer's field. After remedying the trouble we took off and arrived in Edmonton, chilled to the bone, after a three-hour flight.

In Edmonton our business began to take shape. Lin and I found a housekeeping room on 107th Avenue, at a reasonable rent, while Ben Metcalfe took his room at the Royal George Hotel, calling it our traffic office. Our airplane was on wheels at the Edmonton Municipal Airport. Thus, my first job was to fit it with skis for northern operations.

The skis, which had been shipped from Sydney, were on hand and after much work I managed to get the wheels off and the skis on. Everything seemed fine, but unknown to me, the safety cables on the skis, which prevented the ski from nosing down in the slipstream should the rubber shock (bungee) cables break or stretch, were too long.

On January 27, I took off to test the skis. While manoeuvring, I felt a shock through the machine and, looking over the side of the cockpit, saw that the left ski was only prevented from becoming vertical by the safety cable.

My airspeed had fallen off with the drag from the ski so I worked my way around to the downwind side of the aerodrome and headed for the ground. I hoped to make sure that the left ski did not dig in and flip me over on landing. I pulled the nose high just before touch-down, and as the slipstream fell off the ski rose a little, so that, with the high nose position, the ski did not catch and all was well. It was a bit too close for comfort, however. I fitted new bungee cords and shortened the safety cables before trying any further flying.

Our gasoline caches were now in position as far north as Fort Rae, and Ben Metcalfe was hard at work trying to get some business. A well known trader of the north, Henry Swanson, was in town purchasing a load of supplies with which to open a small store at Cameron Bay on Great Bear Lake, near the Arctic Circle. He did not have much money to spare so we arranged a special rate for a full aircraft load because, in addition to getting Swanson's business, we felt that the sooner we made a trip as far north as Great Bear Lake, the better it would look for us. Once people knew we had made such a flight successfully, they would have more confidence in us. We arranged to take Swanson's load north on or about February 10.

I decided to check the operation of my modified ski cables and also some engine winterization work I had done, so I flew a test flight on February 5, with ground temperatures about -30°F. The skis and engine worked well but I froze part of my face in the icy slipstream around my open cockpit. It was apparent that I was going to require a face mask for very cold temperatures. I made one up out of some chamois leather, which, I found later, worked fairly well except that it tended to freeze to my face, requiring careful thawing out before trying to take it off after landing.

Ground temperatures were running from -40° to -50°F throughout the far north at this time so I knew that the trip was not going to be a picnic. No one had flown that far north in an open cockpit in the middle of winter in that area before, and I am sure that no one has done it since then. All had used closed cockpits.

We left Edmonton, heavily loaded with trade goods, plus Henry Swanson and Ben Metcalfe. It was bitterly cold but my face mask worked well. We landed at Fort McMurray and remained overnight. From here on, each morning I would have to be up very early to get my engine heated up and be ready to go shortly after daylight. This meant starting work at least two or three hours before take-off. The next day, February 11, we landed at Fort Chipewyan and could not proceed further due to weather. On the morning of February 13 the weather cleared and we took off, landing at Fort Fitzgerald and Fort Smith and remaining overnight at Fort Resolution. I deliberately landed at each point en route to become familiar with their landing strips on the ice. Everything worked satisfactorily except for the thawing out near a stove in order to remove my face mask. On the 14th we took off but had to return to Resolution due to wing icing. On the 15th we landed at Fort Rae, where the RCMP barracks served as shelter.

Fort Rae was our last gas cache so it was necessary to carry some extra fuel in cans in the cabin on the next leg to Great Bear Lake. I had to ensure that I had enough fuel to get back to Fort Rae from Great Bear Lake. This was going to give us a pretty heavy overload for take off.

Shortly after we had landed, a Civil Aviation Division aircraft arrived, piloted by my old friend Inspector Howard Ingram. I was very surprised to see him as we were not aware that he was in that part of the country.

He strode over and said hello, then asked me to walk over to my aircraft where he looked at my propeller, which was one of the laminated wooden ones. He examined the laminations near the hub and pointed out that the glue had dried and was coming out of some of them. I said that it had started to show in the extreme cold of the Northwest Territories after coming from the damp climate of the East Coast. I said I had planned to fit a new metal propeller on my return, which was also true. I had not had the money to buy one before this. He understood, but warned me that the metal prop must be fitted without fail. He cautioned me that I would be hearing from Ottawa on the subject.

At this time I did not realize what a stern view Ottawa would take of the matter, but I did wonder why Ingram had followed me to Fort Rae to condemn my propeller. I learned years later that my competitors had complained about my prop and as a result the Civil Aviation Division was forced to take action against me. Later on I flew with the people who actually laid the complaint and we all had a chuckle about it. They were perfectly right in regard to my propeller except that it was tough luck that I could not have completed the trip and secured the new prop without trouble.

On February 18 the weather was satisfactory for the Great Bear Lake flight. We loaded the extra fuel in the cabin and I managed to stagger off. It was a very cold day and when we landed at Cameron Bay 3 hours and 30 minutes later, it was -55° on the ground. We had delivered Henry Swanson and his load as agreed, so he paid us in cold, welcome cash.

We picked up a partial load for the south, put our extra fuel in the tanks and left on February 21, arriving back in Edmonton on the 24th. On March 11 my new propeller arrived, was fitted and air tested.

We had arranged a contract with the Hudson's Bay Company to transport trading supplies to their posts as far north as Fort Smith and to bring out fur bales on the return trips. This was a good operation except that there was little profit margin in it.

I found I could not compete with both the flying and the ground work to look after so I hired an air engineer, Bill Sutherland, to help me out. Bill was a genial fellow and a good engineer. He had been a railroad man originally but had lost a leg in an accident. With his artificial leg he did a grand job of getting around in the deep snow and the cold, and always remained cheerful. On some trips, if the load permitted, I would take him with me; otherwise he stayed at base in Edmonton or Fort McMurray.

I flew every day the weather permitted, heavily loaded in both directions, from March 13 until March 30 when the contract expired. At this time I received a terrific shock. A letter arrived from Ottawa about my propeller inspection at Fort Rae. It stated that my pilot's and air engineer's licences were suspended from February 16 until May 16. The Department had been kind in delaying the arrival of the letter, which let me finish my Hudsons' Bay contract, but it still meant that I was out of business for a while. Ben Metcalfe returned to Lethbridge and I had to let Bill Sutherland go, although we were to fly together again later on.

The returns to the company in the East had been so slim, even though George and I drew little salary, that in view of my grounding they decided to go out of business. My aircraft, G-CAHJ, and George's, G-CAHE, were purchased quickly by Grant McConachie who immediately put them to work.

The settlements finalized themselves in a court case in which the oil company sued for the gasoline which had been placed in the north and which was not fully paid for. The judgement wound up with the company paying part of the amount and I, who had authorized and signed for the fuel operation, paying the rest. Lin and I managed to pay our share off over an extended period of time.

After all our work and worry we had finished up without our aircraft, no finances but owing money, grounded, and living in one housekeeping room in Edmonton. An incident occurred at this time which caused me some bother from time to time later on. One day, in our little room, I suddenly grew faint and collapsed, but soon came to. I saw a doctor, but one who was not a pilots' medical examiner, as I did not want to be medically grounded. The doctor checked me over and gave me some pills and told me not to worry. I never mentioned this to our aviation doctors throughout my years of flying, but did so just as I was going to retire on reaching the maximum age.

8

The Jasper Desperados

We lived in our confined quarters in Edmonton until the fall of 1933. The suspension of my pilot's and engineer's licences expired on May 16, but try as I did, there were no jobs to be had in or out of aviation. There were unemployed men lined up for every possible opening. Our small savings were running low. We had many good friends, however, who helped us to survive this difficult period.

Lin and I walked around the city a good deal during these days, and often went to a movie to pass the time. Some of the movie houses were owned by people I had flown to the Peace River country in Grant McConachie's aircraft in November 1932, and for us they never charged admission.

The Department of Transport advised us that from now on all commercial pilots wishing to instruct must have a special certificate. Until now all that was needed was the standard commercial licence. We were advised that F/L Elmer Fullerton of the RCAF would tour Alberta to carry out the tests. I made myself available at Edmonton airport for checkout on July 18, 1933, and was concerned because I had not flown for some time. However, I managed to pass the test and was awarded an instructor's rating. The aircraft used for the tests was an RCAF Gipsy Moth.

In September 1933, through the good offices of Grant McConachie, I was asked to fly a Puss Moth, CF-APE, carrying Prince Galitzine, a Russian nobleman who had escaped from Russia at the time of the Revolution. I was to take him to Jasper for the festivities connected with the annual golf matches. His wife, the Princess Galitzine, was already at Jasper Lodge. I was to have the services of an air engineer, who turned out to be my old friend with the artificial leg, Bill Sutherland. It was good to fly with him again.

I test flew 'APE on September 4 and found it pleasant to fly. It was a small three-place monoplane with folding wings. The pilot sat in the single front seat, with two passenger seats immediately behind, all three being in the enclosed cabin. It also had an unusual yet effective system of air brake: on pulling a small lever in the cockpit, the fairings on the two long undercarriage struts would turn at right angles from their streamlined positions. As they were fairly deep fairings, they presented a goodly surface to the aircraft slipstream, thereby reducing the forward speed.

Bill and I took off from Edmonton on September 9 for Edson, Alberta, where the Prince had his home. We found a suitable place to land, picked up our nabob and took off. The Jasper area is inside the Rocky Mountain chain and the town and lodge were encircled by mountains. At that time there was no airport and, as I was flying on wheels, I would have to find a suitable landing place.

On arrival I looked everything over and could find only one place which would do, and that was on one of the golf fairways at Henry House, a few miles out of Jasper. I made a low pass to chase off some golfers but they did not get the hint. I circled and came in to land. This time the players scattered, enabling me to land and taxi to one side. (In driving through the area by car in 1966 we stopped to look at the old landing area. It now has a proper windsock and is a fairly good grass airport.)

We were picked up by taxi and taken to town, where Bill and I dropped off at a hotel and the Prince continued on to the lodge. He invited us to come over after dinner to join them at the golf dance.

We had a late dinner and then drove out to Jasper Lodge in a car borrowed from a friend of Bill's. Bill drove well in spite of his artificial leg. We parked the car, then joined the Galitzines. When it was time to go back to town we walked to the parking lot accompanied by the Prince. As we opened the car doors we saw that a young man and woman were seated in the rear seat.

The Prince asked them what they were doing there. They replied that they must have got into the wrong car, but that they would appreciate a lift into town. We agreed to take them. Bill got behind the wheel while I took the front passenger seat and, after bidding the Prince goodnight, we drove off.

When we were nicely along the twisting road towards Jasper, I suddenly felt something ram me in the back. Bill, alongside me, also stiffened. This was a hold-up, we were told, and there was a gun covering each of us. We were to keep on driving to town and if we made a false move we would be shot. The couple then told us that they had been working at the lodge and that during the evening they had robbed a number of rooms and were escaping. They had learned of the arrival of the aircraft earlier in the day and had decided to use it for an escape across the border. When we showed up at the lodge they decided to get into our car.

They demanded that I take off immediately for the United States, but I told them that it was a pitch dark night and that I had no lights of any kind on the aircraft, and also that a take-off at that time, among the high mountains, was suicide. I also told them that since I had done a bit of drinking during the evening it would be wise to have something to eat, possibly with some strong coffee. They agreed—we were directed to drive to an all-night Chinese café in the town, and when we arrived we parked the car in front. There wasn't a soul on the street.

The man checked the café then herded us in. They had their guns hidden in their pockets, but "out of sight" did not mean "out of mind." They showed us the guns before the proprietor came to take our orders, and told us they would shoot from under the table if we tried anything.

To put it mildly, we were both terrified, particularly when we saw them in the light. Their eyes were like the eyes of snakes, so I judged that they were both under the influence of dope and that they probably would shoot us without much hesitation if the need arose.

The waiter took our orders of sandwiches and coffee, but when they arrived we both found it almost impossible to swallow as our mouths were so dry. I toyed with the old movie stunt of throwing my hot coffee in their faces but gave it up as in real life it didn't appear too practical.

Things were looking grim when into the restaurant walked our friend the railroad man who had loaned us his car. He had seen it parked outside and had come in to say hello. He stood at the end of the booth and chatted with us. We hoped, very deeply, that he would realize there was something wrong as we made no attempt to introduce the young couple, which we couldn't do in any case as we did not know their names. We also hoped he would decide to take his car, but he didn't. He bade us goodnight and left the café. We waited for something to happen, hoping he had realized we were in trouble, but the police did not arrive. Our friend had noticed nothing wrong.

As it was still dark I said I needed a short rest before flying, and strange as it may seem, our abductors agreed. We left the restaurant and walked the short distance to the hotel, leaving the car where it was. The hotel clerk was not at the desk, so we were marched up to our double room where we were told to rest until daylight.

The couple stayed outside our room, walking up and down the corridor, while Bill and I discussed our situation in whispers. There were no telephones in the rooms and there was a straight drop of 25 feet from our window to the ground. The only course of action was somehow to get to the night clerk and have him call the police.

We waited for a while until there was silence out in the corridor and then we flipped a coin to see who should try to get downstairs first. Bill lost the toss and was, therefore, to make the first attempt. I quietly slipped him out the door and he disappeared down the corridor, but in a moment he was back again, shepherded by the man and woman and their guns. We sat around for a while, then decided to try again. It was automatically my turn this time, so I slipped out and tiptoed down the corridor, expecting to be jumped at the corner of the hall, but there was no sign of anyone. I got downstairs and to the clerk's desk, told him what had happened, and he called the police.

Just then a young man, also a hotel guest, came into the lobby from outside. His face was as white as a ghost. He said he had returned to the hotel a short time before and had encountered the man and woman, who at gun point had forced him to drive them to Henry House just to see where the aircraft was parked. On return to town they had got out of his car near the hotel and let him go. Clearly they felt that we would not try to get out of our room the second time and that they should make sure just where the aircraft was before coming back for us.

The police arrived, rounded up the couple after a brief search and clapped them in jail. We gave our statement to the police and were allowed to go about our business. We rested that day and then took off for Edmonton on September 11. Lin was staggered to hear of our experiences in Jasper, the most unnerving non-flying experience in my life.

9

A Steady Job at Last

The Depression was now at its worst, from the standpoint of aviation. The RCAF was releasing dozens of pilots as an economy measure, and small firms were being crowded out of the business. I was lucky, though, for while I was often waiting for something to turn up, the waiting was never so long that Lin and I faced destitution. Nevertheless there was a fair amount of hopping from job to job.

Grant McConachie asked me to undertake a regular passenger service between Edmonton and Calgary, using the Puss Moth. We tried it in September 1933, making three trips, but there just wasn't enough business. I flew a few times more for Grant up to the end of October, and then no more. We didn't eat heartily in those days, but we seemed to eat happily, taking many more picnics with friends like Pat and Elsie Solway (with whom we were now living in Edmonton) than in the brighter days to come, when steady employment was available.

Lin and I parted briefly in November when I accepted a job as instructor for the Aero Club at Brandon, Manitoba. She went to Medicine Hat to live with her family. My flying club days must have been the absolute nadir of my career—I had one airplane (an Avro Avion), few students and a very cold two-mile walk to the airport every flying day. Perhaps it was significant that across from the airfield was a huge mental hospital. The whole world was an asylum in those days—and a few corners of the world, soon to intrude into our headlines, were basket cases.

Out of the blue, on January 21, 1934, I received a telephone call from Winnipeg. It was G.A. "Tommy" Thompson, assistant general manager of Canadian Airways Ltd. It was the largest flying outfit in Canada, one of the largest in the world, a bush plane corporate empire that extended from coast to coast. Would I care to come to Winnipeg to discuss a possible job with the company? Would I! The possibility of a good, steady salary was almost too good to be true. Within hours I had resigned my position with the flying club and was on the train to Winnipeg.

Early the next day I appeared at Thompson's office. There were two other people also waiting, pilot M.B. "Jock" Barclay, recently of the RCAF and Manitoba Government Air Services, and A.M. "Mickey" Sutherland, a first-class air engineer.

I was called in first. Tommy questioned me in great detail on my aviation background, then told me that I was accepted. He said that two pilot positions were open, one at For McMurray, Alberta, and the other one to operate from Winnipeg into northern Manitoba. I said I was particularly anxious to get back into the Northwest Territories if possible. I was told to wait while he interviewed Jock, after which he said that I was to go to Fort McMurray and Jock was to operate in northern Manitoba. We were both very pleased. Tommy then interviewed Mickey Sutherland, who was also accepted, and was teamed up with me as my engineer.

We went out and did some quiet celebrating that afternoon. To be back on a good salary again was almost unbelievable. I was to receive a basic pay, which as a junior pilot was about $175 a month, plus three cents a mile flying pay. We usually averaged about $375 to $400 a month during steady flying periods—very good money at that time.

I reported, with Mickey, to the Brandon Avenue shops—the major overhaul shops for Canadian Airways, on the banks of the Red River—on January 27, 1934, to test fly my airplane, which had just received a major overhaul. As our planes were nearly all on floats or skis, they could be flown from the water or the ice right in the middle of the city. Mickey and I met T.W. "Tommy" Siers, chief engineer of the whole of Canadian Airways. Tommy showed me my plane, a Fairchild 71, CF-AKY, powered by a 425 hp Pratt and Whitney Wasp C engine. It was sitting on the ice on skis, shining and clean after its overhaul. My heart was in my mouth as I had never flown in an airplane anywhere near the size or power of this one. The maximum power I had handled previously was only 200. I didn't want to say too much, though, for fear of jeopardizing my new job.

Tommy told me to get into the pilot's seat. He leaned over my shoulder and showed me where the various items were that I should know about. He said I was to take off and fly the aircraft over to Stevenson Field, and report on its flying state. Mickey was to go with Tommy by car to meet me at the airport.

I was unfamiliar with the area and had never seen Stevenson Field at all. Tommy pointed the general direction, then got out of the aircraft. I managed to get started and warmed up, but a light snow was starting to fall. Tommy was standing nearby waiting to see me take off so I taxied out on the ice, headed into wind and gave her the gun. I was quite tense until I found that the Fairchild came off nicely and that the controls seemed fine. As Brandon Avenue disappeared behind me I found myself in heavier snow with decreasing visibility. I set course but the snow got thicker and I was forced to do some turning to avoid the worst areas. Finally, when I figured I should be near Stevenson Field, I found I couldn't see much of anything and absolutely nothing that looked like an airfield. Circling, I realized that I was lost on my first flight with my new company, and that a trip of less than five miles. I was almost sick when I thought of what would happen to my new job.

An open field with a good ground cover of snow on it appeared below so I decided to land, as Tommy Siers had warned me that the fuel wouldn't last too long. I did my best with my unfamiliar aircraft and managed a fairly decent landing. There was a house a little distance away so, leaving my engine idling because of the cold, I hurried through the snow. A surprised man came to the door and I asked him if he knew where Stevenson airfield was. He said that on clear nights he could see a rotating beacon flashing to the north. This was the information I needed.

I rushed back to my aircraft and took off, keeping my fingers crossed over the fuel supply, and headed north. In a few minutes I saw the field and landed. Waiting were not only Tommy Siers and Mickey Sutherland, but also the assistant general manager, Tommy Thompson. They were all concerned as I was much overdue; in fact, I think they thought I had cracked up the newly overhauled aircraft.

Tommy Thompson was quite angry with me for getting lost, but when I explained that I had been caught in the snow storm and had force landed in a field, he calmed down and said that I must be able to handle the aircraft if I had done that successfully on my first flight. He said that I was still on the the strength of the company—a great relief indeed. The following day I carried out a test flight from Stevenson Field and found everything working well.

On the morning of January 30, 1934, I took off, bright and early, with Mickey in the cabin and full fuel and oil in the tanks. After 5 hours and 15 minutes of bucking westerly winds, we landed at Regina, where we remained overnight. We took off again the following morning and flew another 5 hours and 15 minutes to Edmonton.

There we reported to district headquarters and our superintendent, C.H. "Punch" Dickins, the famous northern pilot and winner of the McKee Trans-Canada Trophy for 1928. He welcomed us to his staff and told us we would be stationed at Fort McMurray under the orders of the chief pilot of the base, Walter E. Gilbert. We were to load our aircraft fully and depart for Wabiskaw and Chipewyan Lake, where we were to deliver our load, then to fly on to Fort McMurray.

We left on February 2. I was able to get the feel of my aircraft fully loaded—it was a good plane and I enjoyed every minute of flying it. We found the trading posts, delivered our loads and later landed at Fort McMurray.

Here we met Walter Gilbert, one of Canada's finest Arctic pilots and winner of the McKee Trophy for 1933. He introduced us to his staff, including Fred Lundy, the chief traffic and freight officer, and Fred's assistant, Benny Benzie. During the next short while, we met our fellow pilots and engineers. At that time the pilots were Walter Gilbert, W.R. "Wop" May (McKee Trophy winner for 1929), C.M.G. "Con" Farrell, A.M. "Archie" McMullen and myself. The engineers were Don Goodwin, S.A. "Sammy" Tomlinson, Lew Parmenter, Casey Vanderlinden, Al Parker, Rudy Huess, Fred Little, Frank Kelly, Tom Caddick and Mickey Sutherland. Mickey, of course, rode with me as my engineer in CF-AKY.

A new company had entered the north after my old Explorers Air Transport went out of business. It was called Mackenzie Air Services, headed by Leigh Brintnell, and was now our opposition in the area. On the flying staff, in addition to Leigh Brintnell, were three top northern pilots—A.M. "Matt" Berry (McKee Trophy winner for 1936), S.R. "Stan" McMillan and, later, H. Marlowe Kennedy.

We flew every day that the weather permitted and when back in Fort McMurray stayed at the old Franklin Hotel. When possible, I spent time looking for a place for Lin and me to live in. It was a small bush town with few places available but I finally found a small house that had originally been a shop. A little fixing up inside and some furniture would do the trick. I managed to get the bare essentials of furniture together, then asked Lin to join me. She came on the weekly train from Edmonton in the company of Dr. and Mrs. Malcolm McCallum. Mac McCallum was coming in to start his first medical practice and was bringing his bride, Grace, with him. Grace and Mac became our firm friends.

As there was little furniture available in McMurray, I made a few pieces of rough furniture whenever I happened to be home in the evening—a table, a bookcase, a washstand—which were all a bit shaky but did the job. The heating was done by the kitchen coal stove and an "airtight" wood-burning heater in the small living room. The lighting was by coal oil lamps or Coleman gasoline lamps and the water supply was a barrel in the kitchen, filled periodically from a water truck. We were quite happy in our house and were in the same position in regard to utilities as everyone else in the community, so the lack of this or that did not matter. Lin soon made it a real home.

Flying business was good, partly due to new mineral strikes made in the north. We flew from first thing in the morning until almost dark on every suitable day. Our operations were mainly up and down the route to Great Bear Lake or along the route to Aklavik at the mouth of the Mackenzie River. The main staging points on the McMurray-Great Bear Lake run were Fort Chipewyan, Fort Fitzgerald, Fort Smith, Fort Resolution, Fort Rae, then to Cameron Bay on Great Bear Lake. If we were going to Aklavik, we turned off at Fort Resolution and flew along the south shore of Great Slave Lake and then to the north along the Mackenzie River. The main staging points on this route, after leaving Fort Resolution, were Hay River, Fort Providence, Fort Simpson, Wrigley, Fort Norman, Norman Wells, Fort Good Hope, Arctic Red River, Fort McPherson, then Aklavik. From time to time we flew to Coppermine on Coronation Gulf or into the Barren Lands west of Hudson Bay.

Once we had left Fort McMurray we had no base organization to help us and we were more or less on our own. We would be up well before dawn each morning, as in winter it was necessary to spend an hour or two heating the engine and its oil before starting up. The engineer usually stayed with the fire pots under the long engine cover in order to heat the engine, while the pilot used another fire pot to heat the oil which had been drained from the engine the night before. When both oil and engine were good and hot, we filled the oil tank and started the engine as quickly as possible. A few years later a system of oil dilution was introduced, largely by the efforts of Tommy Siers, our chief of maintenance, which eliminated draining the oil the night before. He received the McKee Trophy for 1940.

Mickey and I spent most of our time in the spring of 1934 on the Great Bear Lake run, until about March 23 when we were detailed for our first Barren Lands job. This was usually a tough assignment as this was a huge treeless area of tundra and lakes, not well mapped, and where magnetic compasses were unreliable due to the nearness of the magnetic pole. It was a winter operation, as spring break-up did not occur until May-June. We were to transport a party of prospectors of the Cyril Knight Prospecting Co., headed by Alec H. Stewart, with tons of supplies, to Windy Lake. This lake, near Neultin Lakes, was the site of an Eskimo trading post operated by Dick Halcrow. From there the prospectors were going to work afield, mostly after break-up. The bulk of supplies for the operation had been taken to Stony Rapids post at the eastern end of Lake Athabasca by boat the summer before.

We took off from Fort McMurray for Stony Rapids on March 23 with Stewart and his men plus their personal gear. Before we could move the men and freight from Stony Rapids, we had to put in a gas cache near the midway point. We selected Kasba Lake, as it was just nicely on the edge of the Barrens. There were sufficient stunted trees for fuel and so on for our camp. I made four trips from Stony Rapids, depositing several 45-gallon drums of aviation fuel at Kasba. We then flew in tents and equipment, plus the prospectors, to set up camp. When that was done Mickey and I moved into the camp with our own tent. We also had flown in a team of husky dogs and a sled for the prospectors. The temperature varied from -10° to -50°F during our operation.

On one of the return flights from Kasba to Stony Rapids, we landed on Selwyn Lake and shot two caribou for the RCMP boys to give to the Indians at Stony Rapids, who were out of meat. The RCMP loaned me one of their rifles to do the shooting. Cutting the caribous' throats then loading them into the aircraft was a messy job but we managed. The Indians cleaned up the aircraft afterwards as their part of the arrangement.

From there on we moved freight into Kasba and, weather permitting, on to Windy Lake day after day until finally we landed at Windy with our last load. Alec Stewart was very satisfied with the way we had transported his prospectors and insisted on sending a note to Canadian Airways, stating that we had done a good job for him, and in due course Tommy Thompson complimented us on behalf of the firm.

On the return we were getting close to Kasba when we ran into a real Barren Lands blizzard. I managed to land at our camp, where Mickey and I had to get the oil out of our aircraft fast before it froze, then dig two bridges in the ice to anchor our aircraft to, so that it would not blow away. This is rough work in that kind of weather. That night we crawled into our tent but it was so cold we could not sleep. The temperature was -50° to -55°F and the wind must have been in the range of 40 to 50 mph. Before daylight we got out of our sleeping bags, shaking like leaves, and crawled under the engine cover, where we lighted our fire pots to try to warm both the engine and ourselves. We had not eaten since noon the day before, as we found that we could not prepare food outside in those weather conditions, and we couldn't operate our fire pots in our tent, as we nearly smothered ourselves, so we went hungry. By daylight the wind dropped and the engine and oil were warm, so we took off for Stony Rapids.

We were thawed out and fed by our friends, the two RCMP constables at the post with whom we normally stayed when overnighting. We landed at Fort McMurray later that day, April 8, where we stayed for three days getting our gear in order, as we had been told we were to go north and operate in the Great Bear Lake area over the break-up period. Wop May and his engineer, Casey Vanderlinden, were to do the same thing at Aklavik. In both cases this meant working further and further north as the ice melted, until finally we would have to pull our machines up on the beach at Cameron Bay and Aklavik respectively. We would change them over to floats before we could fly south again. The floats had been taken in by boat the previous summer.

I said goodbye to Lin on April 11, and Mickey and I took off from Fort McMurray, not to return until June 10. We were now flying in Fairchild 71 CF-ATZ, an airplane in which I was to have many adventures.

10

The Blasting Brewers of Great Bear Lake

We flew north for Great Bear Lake fully loaded, including extra clothing and necessities for ourselves. My kit included some malt extract for making beer, as I knew there wouldn't be much available at our destination.

The signs of spring break-up were appearing on leaving Fort McMurray and as far north as Fort Fitzgerald and Fort Smith, but thereafter winter was still in control. Beyond Fort Rae we had to make a landing at Hottah Lake to deliver part of our load. A new uranium strike had just been made there by Ted Hargreaves and his father, an eastern mining and prospecting family. We then flew into Cameron Bay and checked in with Mr. and Mrs. Harry Reed in the inn. They were a Spokane, Washington, couple who had come north and built a large log cabin in which they served food and rented floor space for sleeping bags in the main living and dining room. Mickey and I were allotted our patches of floor, which we would increasingly use as the spring break-up progressed.

There was a shortage of food at Cameron Bay, as part of the year's supplies had been lost being transported across Great Bear Lake from Fort Franklin the previous fall. The tug *Speed* had caught fire and sunk in heavy weather, and the barges had drifted ashore and been wrecked. As a result, a famous northerner, Vic Ingraham, lost both his legs, but by sheer courage returned north later and carved out a good business for himself.

We continued to fly almost every day, sometimes in the general Great Bear Lake area to such places as Contact Lake, Spark Plug Lake, Hottah Lake and McVicar Arm, servicing the mining crews which were scattered throughout the north and required supplies or transfers of personnel. Occasionally our loads took us as far south as we could go on skis which, by April 20, was Fort Fitzgerald. Very shortly after our last take-off from Fitzgerald the ice on the Slave River went out with a roar. We then only came as far south as Fort Resolution on Great Slave Lake. Eventually ice conditions became unsatisfactory even at Fort Resolution and Fort Rae so that our flying operations shrank to the general area of Great Bear Lake or north of there.

Food was becoming a problem at Cameron Bay, with its population of 25 or 30 people. One day late in May I was asked to fly north along the shore of Great Bear Lake to an uninhabited cabin in which hunters late in the previous fall had stored a lot of caribou legs after a successful hunt. Mickey and I were to bring a load of this meat back to Cameron Bay. We found the cabin easily enough, but the meat was getting quite high, as there had been a warm spell shortly after it had been placed there. We loaded our aircraft and flew to Cameron Bay, but the odour of the meat made us both ill en route.

The most suitable parts of the meat were eaten by the inhabitants, including ourselves, resulting in everyone getting a bad case of diarrhea. The outside toilets and places in the stunted bush nearby were in great demand most of the time. Flying became a veritable gamble—one is tempted to use a rude simile—for the toilet facilities in the aircraft were nil. We had no doctor and very few drugs so we just had to make the best of it.

At Cameron Bay Canadian Airways had a tiny cabin for wireless equipment used to handle company traffic and weather messages. We had a similar post at Fort Rae. The wireless operators were Henry Roth at Fort Rae and Jack Green at Cameron Bay. The stations passed traffic messages back and forth but could not be used by our aircraft as there was no wireless equipment in the planes.

In addition to the company wireless stations there were stations of the Royal Canadian Corps of Signals, Canadian Army, at many points. They also passed traffic messages of many kinds. At Cameron Bay the RCCS station was operated by Cpl. Frank Rapp. (During the Second World War and later he became S/L Rapp of the RCAF, another friend.)

I mentioned one day to Frank that I wanted to start a brew of beer with the materials I had brought in. Frank was interested and suggested that if we started the brew in his cabin, he would find a suitable crock. We didn't like to start the process in our company wireless cabin because Jack Green was already making his own beer there and we thought that one brew per cabin was enough.

Frank, Mickey Sutherland and I started the brew and on occasions when we were not in the air we would meet like a gang of conspirators, pull the crock from beneath Frank Rapp's bed and stir it. Once the RCMP sergeant opened the door and walked in. He took one look and started to laugh and from then on helped us stir the brew. Mickey did not drink alcohol at all but took his full part in all the preparations.

We were doing a lot of flying to various mining and prospecting camps while the ice conditions were suitable, although we were getting weaker as a result of the diarrhea. About the end of May we had finished all our operations so we pulled the aircraft off the ice and onto the small piece of flat beach. I had made arrangements with Eldorado Mines, a few miles away, to borrow a few cases of dynamite so we picked this up just before pulling the machine on shore. The ice in Cameron Bay was still about two feet thick but I figured that by exploding the dynamite under water at many points, in the shape of a landing strip, I could hasten our departure for the south on floats.

Using ice chisels, we dug holes down through the ice. We would then put a few sticks of dynamite in an old burlap bag or anything else suitable, properly fused for underwater operations, tying the bag to a stick. We would lower the bag beneath the ice, letting it hang from a stick placed across the top of the hole, light the fuse and get off the ice. It took a few shots to arrive at the proper number of sticks of dynamite to really open it up. Everything was going well until one sack of dynamite did not explode. We waited ashore for quite a while, then Mickey and I went out and slowly raised the sack. We could hardly breathe as we opened it up, but we found that the fuse had gone out. We replaced it and it fired OK.

The breaking up of the ice caused it to melt much faster, and by using boats, manned by willing helpers, we gradually kept the chunks of ice moving over to one side. Finally a short strip of open water started to show.

While these operations were going on we decided to sample our beer. It was still a bit on the green side but we were in the mood to try it. We invited our friends and helpers in and soon demolished the beer. From there on things happened. A baseball and bat were produced and somebody organized a cockeyed sort of baseball game among the rocks immediately above the camp. I sprained a finger trying to catch a ball that I couldn't see properly due to the overload of beer. The game terminated as more assorted bottles of alcohol, which had been hidden away by various individuals, were produced. Everybody except Mickey and one or two others celebrated, until at length we crawled into our sleeping bags and passed out.

In the next few days we managed to replace our ski undercarriage with floats. The ice was melting and there was sufficient open water for a take-off "light," that is, without a load. It would be close to the limit but I figured it could be done.

The afternoon before our take-off the first aircraft arrived from the south on floats. It was Stan McMillan of Mackenzie Air Services. He looked over the water strip from the air and decided it was a bit short for a landing with a load. He touched his floats down on the ice, slid along and dropped into the water without doing any damage—a beautiful bit of flying. It was good to see him. Aboard his aircraft, among other things, was some extract of wild strawberry for our Cameron Bay diarrhea, and what a scramble there was for that!

The next day, June 10, 1934, we slid our seaplane into the water with only enough fuel in the tanks to get to the White Eagle Silver Mine on the Camsell River, about 40 or 50 miles to the south. This was the nearest really open water. Mickey and I crawled aboard and warmed up the engine, then I taxied to the very edge of the ice, downwind, turned into wind and opened the throttle wide. I got her quickly on the "the step" and, as the edge of the ice was almost under our keels, gave the elevators a real heave. She came out, barely touching the ice's edge with her keels as we passed over. We both breathed a sigh of relief, then flew over to the Camsell River where we landed and refuelled. We picked up some goods for the south and took off for Fort Rae. The next day we made load stops at Rocher River, Fort Fitzgerald, then settled on the water at Fort McMurray. Home again, much thinner, after two months away.

* * *

My Fairchild, CF-ATZ, required a fair amount of repair and, as Wop May was on leave, Walter Gilbert instructed me to take his aircraft, Bellanca Pacemaker, CF-AKI, and his engineer, Casey Vanderlinden, and go north with it. Casey and I took off with a full load on June 14, landing at various posts and spending the night at Fort Rae. From Rae we flew north the next day into Hottah Lake with its new uranium find. From here we were to pick up a crew of prospectors on an unnamed lake. We got badly caught by the weather closing in at dusk, just managing to edge over the last rocky hill on the edge of the lake. We landed and had to be guided to shore by lights from the prospectors, with whom we camped for the night. The next morning turned out fine so we transported the prospectors to Cameron Bay. Our water strip there was now in good shape with the continued melting.

On the 17th and 18th we moved loads between Hottah Lake, Fort Rae, Fort Resolution and back to Cameron Bay, where a message awaited us from the White Eagle Silver Mine on Camsell River that one of their men had fallen 40 feet down the rock shaft they were digging. He was in bad shape and urgently required transportation to hospital in Edmonton, a thousand miles distant. We took off immediately for Camsell River, arriving just before dark.

With some pieces of wood we managed to fashion a bunk type of stretcher in the Bellanca, securely fastened to the floor. At first light we carried the injured man into the aircraft and strapped him to the bunk so that he could not be thrown off. The manager of the mine decided to send one of his staff along to help. It turned out the assistant was very much bushed and in need of mental help, but we didn't realize this immediately. The mine manager placed part of the mine medical kit aboard in case we needed it. On top of everything else, as is frequently the case when time is important in an emergency in the bush, the weather was very poor.

We took off as soon as visibility permitted but I realized immediately that it was going to be a dirty trip. It was a case of flying close to the tree tops almost all the way to Fort Rae. As well as flying a general compass course, my map reading had to be dead on, for to get away from my planned route might mean flying into a hill or knoll.

In the Bellanca cockpit, Casey and I sat side by side. While I concentrated on my very low-level map reading, Casey kept an eye on the patient and the helper immediately behind us. Casey nudged me and muttered in my ear that the helper was telling the injured man that he shouldn't worry about his injuries because the weather was so bad that we were going to crash anyhow. This startled us. I told Casey to tell him to cheer the patient up as we would surely get through. But I was worried about the state of mind of the helper. A bit further along Casey asked me to look around. The helper was trying to get the injured man to swallow some pills from the medicine supplies. What kind of pills they were didn't seem to matter. We told the helper not to administer medicine of any kind, and from there on we watched him as carefully as the injured man.

We groped our way into Fort Rae, refuelled and took off, and as we went further south we ran into gradually improving weather. Late in the day we landed at Fort Smith. We stayed overnight, leaving our patient in medical hands. Next morning, June 20, we took off for Fort McMurray. The medical men at Fort Smith told us to get our patient to hospital in Edmonton as soon as possible.

At Fort McMurray I found that Walter Gilbert and Fred Lundy had decided that I should change aircraft as my Fairchild, CF-ATZ, was now in good shape, whereas Bellanca, CF-AKI, needed attention from the engineers. A stretcher bunk had been fitted in the Fairchild for the patient. I was still half sick from trouble with my innards and diarrhea, so I was told to take a rest for a week when I got to Edmonton. I got hold of Lin and told her to pack a bag as she should come along also. One more passenger was put on board in Fort McMurray—Henry Swanson, the trader I had carried into Great Bear Lake during my Explorers Air Transport days. His enterprise in the north had failed and he was on his way out.

The day was extremely hot and after take-off I found that the cabin partly filled with burnt oil fumes which, on top of the smell of the drugs from the injured man, plus the heat, made it very uncomfortable indeed. The air was turbulent and we all had trouble trying not to be ill.

After a rough hot ride of three hours and five minutes, we landed at

Cooking Lake, just east of Edmonton. An ambulance and doctor were waiting and rushed our patient to hospital. Mickey tied our airplane up for a week's stay.

We went into Edmonton and Lin, Mickey, the injured man's helper and I all registered at the MacDonald Hotel. While getting bathed and generally cleaned up I received a visit from the hotel manager who said that the helper was behaving oddly, running around in the corridor with little or nothing on. On my advice, a doctor was summoned, and he was taken away for mental treatment. The patient we had flown in from the mine survived by a whisker; without air evacuation he would have been a goner.

On June 27, with Lin and Mickey and a full cargo load aboard, we left Cooking Lake for Fort McMurray. Home again. On my return I learned that Mickey was to be a base engineer for a while and that my new engineer was to be Frank Kelly. (Frank had been with Canadian Airways for some time. When TCA was formed three years later, he transferred and is now retired from Air Canada.)

11

A Hectic Season

Archie McMullen, the pilot of Bellanca, CF-AIA, and I, with my Fairchild, CF-ATZ, received orders to fly out some loads of pitchblende (uranium oxide) ore on our next flights southbound from the Great Bear Lake area. The ore was to be picked up at the rich uranium workings at Hottah Lake, flown to Fort McMurray, then shipped to Port Hope, Ontario for refining. It would have taken most of the summer to get the ore to Fort McMurray any other way.

A day or two later we picked up our first ore, crushed and bagged, at Hottah Lake. The ore sacks were small but heavy, and it didn't require many sacks to make up a full load. To keep the cargo near the centre of gravity of the aircraft we had to pile them mainly behind my pilot seat. We did not realize in those days the hazard of being close to such a "hot" load. So far as we were concerned, radioactivity was something that made watch dials glow; no one paid much attention to protection while transporting the ore. Frank Kelly actually spread his eiderdown sleeping bag on top of the ore and had a good snooze. God knows how much radiation we absorbed.

Archie and I, with our engineers, made two trips south with uranium ore, sitting with it at least seven or eight flying hours per trip. We all purchased shares in the Hottah Lake development but subsequently lost our investment when it was found that the ore was in a very rich capping on the rock surface, with nothing much at any depth below. It was soon all removed and the mine closed down.

That late summer of 1934 was one of the most varied and eventful in my life. There were a couple of mercy flights (thankfully with no mad flight attendants). Then there were hijinx. On August 2 we had arrived at Cameron Bay, unloaded our aircraft and tied up for the night. Con Farrell and his engineer, Al Parker, flying about an hour behind us, were delivering two live pigs to Harry Reed, the innkeeper at Cameron Bay. We knew that Con had raised the devil about flying the pigs in, but had been overruled. Frank Kelly and I quickly organized the townspeople to gather on the lakeshore with clothes pins on their noses to greet Con on his arrival. This they did, and the ensuing language from Con was a fine display of local profanity. Frank Kelly and I stood back, bursting with laughter. The pigs, named Tillie and Mac, survived at Cameron Bay until about December when they were overcome with carbon monoxide fumes from a heater which Harry Reed had arranged to keep them warm.

Early in August we were southbound from Cameron Bay on Great Bear Lake. On arrival at Fort Chipewyan on Lake Athabasca we received orders to take off our load and exchange it for one to be taken to Stony Rapids, a post at the eastern end of Lake Athabasca. From there we were to undertake another large operation into the Barren Lands, carrying several tons of supplies to the Revillon Freres post at Windy Lake. We were also to carry out some exploration and prospecting flights for the mining company of Cyril Knight Prospecting, old customers by now, with the party again headed by our friend Alec Stewart. This time, instead of a single aircraft and crew, the job was to be done by two machines. The other one was Junkers, CF-AMZ, piloted by Con Farrell and accompanied by his engineer, Al Parker.

Frank Kelly and I joined forces with Con and Al at Stony Rapids, where we all stayed at the RCMP post. We made several flights between Stony Rapids and our old campsite at Kasba Lake to establish a fuel cache there. That done, we took in the Windy Lake post supplies first—Con flew the loads from Stony Rapids to Kasba and I flew them into Windy Lake. We did two or three round trips per day while the weather was flyable, which required five or six flying hours per day per aircraft. The refuelling and loading and unloading also took up a lot of time. In addition we had to service our aircraft, cook our food and look after our camp.

The Cyril Knight prospecting party was looking for gold in the Barren Lands. Stories had circulated from time to time of Eskimos using gold for various purposes, including the casting of balls for their old Hudson's Bay trading muskets. Whether anyone had ever seen any of this gold was doubtful, but the prospectors wanted to have a look around. We flew them in to Windy Lake and left them there until we had tranported most of the supplies in. They wanted to talk to the Windy Lake post's trader, Dick Halcrow, who traded with the Barren Lands Eskimo, so might be able to help them.

After several days of heavy freighting we had a day of bad weather. We needed some fresh meat for our camp. The two engineers were fully occupied servicing the aircraft and Con, no mean chef, had taken charge of the cooking. The hunting, therefore, was my job. The only weapon we carried was a folding gun for each aircraft as part of our emergency kits. The guns, called "Game-Getters," consisted of an over-and-under barrel arrangement on a folding metal stock. One barrel fired .22 calibre ammunition and the other one fired a ball cartridge or a .410 shotgun shell. Their accuracy was poor.

I struck out through the lightly treed country (Kasba Lake is just on the edge of the true barrens) and before I had gone far I scared up a good-sized caribou which had been resting or sleeping in a hollow. It fled from me so I raised my Game-Getter, loaded with a ball cartridge, took aim and fired. The caribou stopped in its tracks, turned right around and charged me. Caribou will not normally do this, but we found later that my bullet had struck its spine near the shoulder so that is was probably unaware of its own actions. As it thundered towards me, I ejected the empty shell, hurriedly placed another in the breech, took aim and fired. This time I got it between the eyes, which is no trick at all when the animal is almost on top of you. It went down, definitely finished, so with my hunting knife I cut its throat (not a very pleasant thing to do) to allow it to bleed properly. We cut it up and carried the meat to our camp. Having no other means of keeping it cool, we staked pieces of it in the icy waters of the lake. We had sufficient meat for our stay at Kasba.

Between flights, Con made some very good stews with the meat and on one occasion said that he had put in a certain part of the caribou's anatomy which is not normally eaten. Con and I were always pulling each other's leg so he said that he had made sure that I ate the undesirable part. I claimed it would be unrecognizable in a stew and that he probably ate it himself. We joked about this for years to come.

On August 19 I picked up some of the Cyril Knight prospectors at Windy Lake along with the trader, Dick Halcrow. We took off and flew in a north-northeast direction until Dick showed us a lake with a small Eskimo encampment on its shores. We landed and talked to the Eskimos, Dick interpreting. They were a bit frightened of the airplane as some of them had never seen one before.

We took off again, making two more landings at lakes having Eskimo encampments, and from the last one got some information about area minerals that was interesting. We took aboard one Eskimo, after much persuasion, to show us the area about which they had talked. We flew again further northeast to somewhere in the neighbourhood of Maguse Lake. It must be remembered that the Barren Lands were poorly mapped at this time and our aircraft compasses were unreliable in this area. It was, therefore, of the utmost importance to keep in mind a picture of the route covered in order to get safely back again. I usually did some sketching of lakes as I flew in, then flew the sketches in reverse coming out. It was completely barren country with many lakes and streams, many with rocky shores. Very easy country to get lost in.

Towards dusk the Eskimo pointed to a lake and said that was our destination. We were somewhere south of Baker Lake, in the general area of Maguse Lake and Padlei post. After looking it over I went in and landed. There were many big rocks underwater which showed up from the air and which I had to remember during and after landing. We planned to remain overnight so I pulled into a small bay which had a gravel beach but which required careful steering around rocks in order to get into it. It was a calm evening so I nosed the aircraft into the beach and we tied the nose and wings to suitable rocks, there being no trees.

We pitched the Mount Logan type of tent which belonged to the prospectors, and Dick Halcrow, Frank Kelly and I prepared supper while the prospectors scouted around in the brief interval before dark. Shortly afterwards we all crowded into the tent and went to sleep, tired out.

I awakened later with the feeling that something was wrong, then realized that it was blowing a gale, with driving rain and snow. Winter sometimes comes early in the Barrens. I knew that the wind was blowing on shore and that my aircraft was tied with its tail to the wind, which is not very good on floats. I yelled to the group to get up and we groped our way through the stormy blackness to the shore.

The aircraft was riding badly with big waves driving right across the tail of the floats. It was impossible to turn it around to nose it into wind in such a storm and darkness, so we used all our extra rope to put another line from the propeller shaft to a big rock, in case the wing lines broke. It then seemed to me that the floats were getting lower in the water at the stern. The aircraft was pounding hard up and down on the gravel bottom, making me fear that the floats might have been holed. We crawled aboard and removed the rest of our ration kits just in case, then stayed up all night pulling on ropes and doing what we could to save the aircraft. By morning we were very tired but the storm had largely spent itself.

The aircraft was sitting with its nose on the gravel beach and the tail of its floats underwater. The tail surfaces of the aircraft were almost on the water. We would have to get the water out of the rear float compartments. We learned later that, although the bottoms of the floats were pounded almost flat, they were not leaking but had taken in water through the vent holes in the rubber caps over each compartment, holes that were fairly large in the old Fairchild floats. (We may have finally removed the caps in order to settle the floats on the bottom.)

We got the aircraft tool kit and float pump ashore, plus some spare hose which Frank Kelly always carried with him. The prospectors, Dick Halcrow and our Eskimo friend got busy carrying rocks into the water on each side of the floats to make ramps to stand on to do some pumping. Frank Kelly devised a lengthened float pump hose by connecting two hoses with clamps and tire tape, all held together with an all-purpose glue which we carried in our kits. He fitted a second hose to it to carry return air to the float. The hoses were then fitted through holes cut in our spare rubber float cap. They also were carefully sealed with tape and glue.

When we were ready one of us got into the icy water, removed a rear float cap and substituted our float cap with its hoses attached. We then started to pump. We pumped in shifts because we found that by the time we emptied one compartment and quickly substitued a carefully sealed cap for our cap with hoses, the compartment was about three quarters filled again. The water rushed in when the cap was changed, no matter how fast we were. Each time there was a little gain, however. Finally, after a day of pumping and changing caps we had raised the rear of the floats to just above the surface. It took most of the next day to finally clear the floats and ensure that they were not leaking appreciably. We had moved our pumping operation from one side of the aircraft to the other as the floats gradually surfaced.

During the evening of the first day's pumping we had our supper cooking in two frying pans. We were all sitting around in a circle in the dark watching the food when I happened to glance away from the pans. I was startled to see eyes shining in the darkness in the light of the fire. I nudged Dick Halcrow who called out in Eskimo. The eyes edged closer. It was a little group of Eskimos who had seen our fire and who now moved in and made themselves at home. In one frying pan we were cooking the last of our dessicated potatoes, having soaked them during the day. One of the Eskimos reached into the hot water, grabbed a very hot piece of potato, stuck it into his mouth, then found it was too hot or that he didn't like it. He simply leaned over and spat it back into the frying pan! He had a very large swelling on his forehead which was causing pus to run down his face. Needless to say we didn't eat those particular dessicated potatoes.

The group of Eskimos stayed with us as long as we were there and helped with the pumping and gathering firewood. The firewood in the Barrens is found in the form of little pieces of old wood in hollows and odd places. It is hard to find but the Eskimos were experts. While we were doing the second period of pumping and preparing to leave, the prospecting group was covering the area looking for minerals. They found some indicators such as greenstone, but no actual minerals.

On the third day we were ready to go again, although in addition to the terrible pounding the floats had taken we found that the exhaust pipe on one side was burning through. Nevertheless, after saying goodbye to our Eskimo friends, we took off and flew back to Windy Lake where I dropped the prospecting crew again. There was some flame coming back from the leaking exhaust system but it seemed to be far enough away from the fuselage fabric. I took off again and flew to Kasba Lake where Con Farrell was still camping. He had been freighting loads to Windy Lake during our absence. We each did four or five more trips to Windy during the next two days; then on the last day I flew the prospectors over to Neultin Lake, then took a load of their gear to Padlei post to the north.

On August 28 we returned to Fort McMurray somewhat battered, bearded and dirty, but intact; and with our assignment complete. Two days later the aircraft was ready to go again with a new exhaust system and general check over. Nothing could be done about new floats at that point so I flew with the flat-bottomed ones until freeze-up later on in northern Manitoba.

My next job was the mail run down the Mackenzie River as far as Fort Norman and return. This involved stops to load and unload freight and passengers at a half-dozen communities—Fort Resolution on Great Slave Lake, Hay River, Fort Providence, Fort Simpson, Wrigley and Fort Norman, then back again, arriving at Fort McMurray on September 2.

Next day we were away again for the Great Bear Lake area, including a flight to Coppermine on the Arctic coast. We made two or three return trips from Fort McMurray to Great Bear Lake until September 20.

On the 21st we started off with a heavy load for Aklavik, at the mouth of the Mackenzie River on the Arctic coast. Beyond Fort Norman and Norman Wells our regular stops were at Fort Good Hope, Arctic Red River, Fort McPherson and then Aklavik. In bad weather we were very careful to keep track of our position going into Aklavik, as the huge Mackenzie delta is a mass of smaller streams, lakes and muskeg in which one could easily go astray.

We returned to Fort McMurray on September 27. The ice was beginning to form around the edges of the lakes and streams in the far north so our work was now going to move gradually further south. On our return to base, Wop May and his engineer, Art Rankin, in Bellanca, CF-AKI, and engineer Frank Kelly and myself in my Fairchild, CF-ATZ, were told to be prepared to depart as soon as possible for northern Manitoba to assist in company operations in that area. The freeze-up in northern Manitoba came later than in the Northwest Territories, so we could do almost another month's flying before changing over to skis.

Wop and Art took off a day or so ahead of us but Frank Kelly and I were ready for departure on October 2. I said goodbye again to Lin—in fact I had seen very little of her all summer. The following day we got as far as the seaplane base at Emma Lake, Saskatchewan, just north of Prince Albert, after 4 hours and 30 minutes in the air. When flying with floats in this area we always flew close to the North Saskatchewan River, in case of a forced landing, there being little other suitable water for seaplanes.

On the 4th we followed the river again to The Pas where we landed and refuelled. Here we were instructed to take a part load to Cormorant Lake, northwest of The Pas, then return. After a day's delay by weather we landed on Playgreen Lake at the north end of Lake Winnipeg where our air freighting base, Norway House, was.

The job was to take heavy freight, pieces of mining machinery, drums of gasoline, dynamite and so on into a new mine on Island Lake, about 150 miles east of Norway House. On this operation we had to leave our engineers at the base in order to carry more freight, but it meant a lot more ground handling of the seaplane and loading and unloading of the aircraft by the pilot. The route itself wasn't a bad run, although from the point of view of a seaplane operation it was a bit dry.

We all stayed at the Playgreen Inn. As well as Wop May and myself, there were pilots Mike de Blicquy, Marlowe Kennedy and one or two others. Marlowe was flying a Junkers with a Junkers Jumo L.5. liquid-cooled engine, CF-AQW. Later on, I believe, the aircraft was converted to a Pratt and Whitney engine. (Marlowe left Canadian Airways in November and joined Mackenzie Air Services.) We all flew at least two round trips per day, weather permitting. In the evenings we got together with a drink or two.

All went well until October 13. I was flying along with a load of gasoline drums and a case or two of dynamite. About three quarters of the distance to Island Lake I got a most peculiar feeling. Coloured spots were dancing in front of my eyes. I shook my head but my vision seemed to get worse. I started to lose altitude as I was approaching the western end of Island Lake. Finally I was down close to the water and not too far from the mine. I was almost unconscious but managed to put the machine on the water. I didn't manage to close the throttle fully and as a result, when I became completely unconscious, one rudder pedal was pushed further than the other and the aircraft taxied around and around in circles. Personnel on shore at the mine got a boat and came out. After some difficulty one of them got aboard my aircraft and closed the throttle. At this point I came to sufficiently to taxi the machine to the dock and have it tied up.

I was taken to the mine hospital and put to bed. The doctor thought it was a case of carbon monoxide poisoning and made me remain in hospital overnight. Next morning engineers were flown in and a leak in my exhaust system was rectified. It had been a close call.

We kept the freighting going full out until the night of October 17 when the ice began to form quickly at Island Lake and Norway House. We decided to head south for Winnipeg and on the morning of October 18 I took off for the south with Frank Kelly and all our gear aboard.

As we headed south we lost sight of the other aircraft as the weather started to close in. After 2 hours and 45 minutes of flying down the lake, I was forced to land at a tiny place called Loon Straits. The weather remained very bad the next day. On the 20th we took off, only to be forced to land 20 minutes later at a settlement named Hole River. In each Icelandic settlement we were taken in, fed and given a place to sleep by the townspeople. They made us welcomed to what little they had. At one of these places we spread our sleeping bags on top of a floor of potatoes, for the room was filled half way to the ceiling with potatoes to keep them from freezing. The next day, the 21st, we made it to the river in front of the Brandon Avenue shops in Winnipeg. There we turned our aircraft over to Tommy Siers and his engineers for overhaul. I took a train out to southern Alberta where I joined Lin, our family and friends and had a long-overdue holiday.

12

Operations at -55°

I returned by train to Winnipeg on November 26, 1934, where I found my Fairchild, CF-ATZ, on wheels and ready to go. The floats that had taken such a beating in the Barren Lands storm had been removed and were standing up against the shop wall, their bottoms looking as though they had been beaten with heavy sledge hammers but miraculously with no leaks in them. They were useless for further operations without a major rebuild.

On November 27 I test flew the aircraft from Stevenson Field. She flew nicely, so Frank Kelly and I loaded our ski equipment aboard and prepared to start for Fort McMurray. This we did the following day, arriving in Regina after 3 hours and 55 minutes. There was a light snow cover on the Regina field, enough for a ski take-off so Frank and I got busy jacking up our aircraft, removing the wheel undercarriage and fitting the skis.

The weather was not good the next morning but we took off along the route to Edmonton, via Saskatoon. The weather deteriorated until we were flying close to the ground, where it appeared that the snow cover was so thin that a forced landing would be difficult. Finally the weather closed right in with little visibility. I tried to turn back to Regina but, trapped by fog and light snow, I had to find a field suitable for a forced landing. I could see little of the terrain below. Swinging over a row of telephone wires I managed to land in what looked like a reasonable field. We didn't slide well on our skis as the snow surface was even lighter than it appeared from the air.

We were in a small, pie-shaped field alongside the railroad track with wires or fences all around it, a tough place from which to take off again. We found that we were close to Davidson, Saskatchewan. Local people drove us into town, where we ate and managed to find out about the weather ahead. It seemed to be improving enough for an afternoon departure.

Later we started up and taxied to the end of the field with some difficulty as the skis were not sliding well on the scattering of snow. After getting into the only possible take-off position, which would mean climbing over rows of telephone wires at the other end of the field, I opened the throttle wide. The take-off was sluggish because of the drag of the skis, but I was committed and had to yank it off the ground.

My speed and altitude were insufficient at this point to climb *over* the telephone wires so I had to go *under* them. I had to line up my aircraft fast to go through the very middle of the gap between the telephone poles, and also be sure that I was high enough not to hook my skis on the wire fence that ran across my course close to the telephone line. I just missed the fence with my skis but felt a crunch and a jar above the cockpit. I had sliced through some of the telephone wires with the tips of the propeller. Turning back to have a look, I saw a number of wires draped on the ground, with people who had watched our take-off waving to us and pointing to the cut wires. My main concern was what damage I had done to my aircraft. It seemed to fly well enough and the engine was not rough, so I decided to try for Saskatoon to the north. The weather was still poor but I managed to feel my way into the Saskatoon field. Frank Kelly and I heaved a great sigh of relief.

On examining our aircraft we found everything correct except for two heavy gouges, one for each ripped-out line, in each of our propeller blades. Frank got busy with a file to smooth the gouges while I phoned to report what had happened. I heard that one of the lines I had cut was one used by the gentleman who owned the major portion of our company, James A. Richardson of Winnipeg! I was going to be a bit unpopular for a while, but at least we had managed to save the aircraft.

We couldn't get out of Saskatoon due to weather until December 2, when we did four hours of flying to reach Edmonton. Here the matter of the telephone lines was reported officially to our superintendent, Punch Dickins, and the aircraft was given a thorough check. On December 6, we flew a full load into Fort McMurray. We were home again for the first time since our departure on October 2.

Lin had returned to Fort McMurray before our arrival so that our house was clean, shiny, warm, and ready for another winter. My brother, J.D. "Dick" Leigh, had arrived in Fort McMurray also. He was interested in becoming an air engineer and was starting work with the company as an engineer's helper to learn the business thoroughly. He took a permanent room at the old Franklin Hotel, operated by Mr. and Mrs. O'Coffey, where, in his spare time, he did a great deal of studying. Dick became a good engineer and in later years was one of the top engineering executives of Air Canada.

The staff of Canadian Airways in the northwest area at this time included Punch Dickins, our superintendent in Edmonton, and the following pilots in Fort McMurray: Walter Gilbert, Wop May, Archie McMullen,

Con Farrell, Rudy Huess (recently promoted from air engineer to pilot after receiving flying training in Edmonton) and myself. The air engineers were Don Goodwin, Lew Parmenter, Frank Hartley, Mickey Sutherland, Art Rankin, Bill Jacquot, Frank Kelly, Sammy Tomlinson, Casey Vanderlinden, Rex Turpenning and my brother, Dick Leigh. Our traffic agents were Fred Lundy and Benny Benzie. Our radio operators at the company stations were now Bill Hartree at Fort McMurray, R.A. Ingrey at Fort Rae and Henry Roth at Cameron Bay.

We had to fly like fiends to catch up on the backlog of work held over from the freeze-up period. Frank Kelly and I in CF-ATZ were sent out with some prospectors, equipment and mail, north along the Great Bear Lake route. The prospectors were Johnny Baker and his three men, who were to be landed in the uninhabited Yellowknife area and picked up again later on. There was then no activity there and our landing was, I believe, the first in the region. Johnny Baker must have known exactly where he was going to operate before we took him in. We landed in the Yellowknife area on December 9, 1934, and left him, his crew and equipment on the ice near a rocky hill. The temperature was -40° to -50°F. We promised to pick them up after Christmas. After completing our flight to Cameron Bay, we turned south to Fort McMurray, arriving there December 13.

The landing area at Fort McMurray was a water channel which ran between the Clearwater and Athabasca rivers, just before their union a little further on. Rudy Huess in Fairchld CF-AAO had alighted as usual on the frozen channel and was turning around at the end of his landing run when his aircraft broke through the ice in a thin spot, presumably caused by a warm spring under the ice. His aircraft settled with its wings on the ice surface and Rudy had to escape by smashing the windshield with a fire extinguisher. The engineers had a long, cold job in getting the aircraft out of the water and back into action.

On the 17th we were sent north with a load along the Mackenzie River as far as Fort Norman. We were also instructed to bring out an old Indian lady who had gone insane, along with an RCMP officer and a lady from Fort Norman who was to act as a matron. On the return journey, the Indian woman was lying in a sleeping bag on the floor of the aircraft. She had pulled the end of the bag over her face. The policeman and the matron, becoming suspicious, pulled the cover from her face to find that she had torn up pieces of cloth and stuffed them up her nose and in her mouth in an attempt to kill herself. The bits of cloth were removed only in the nick of time. From there on she was watched more carefully. We landed at Fort McMurray on Christmas Eve and, due to the condition of the patient, we were instructed to fly her to Edmonton on Christmas Day. We missed our Christmas at home, something which frequently happened in the bush, but on the return flight to Fort McMurray carried a good supply of Christmas liquor for our pilots and engineers on the base. Better late than never.

After a day at home I was sent north again, this time without Frank Kelly. He was replaced on this trip by Sammy Tomlinson, one of our senior base engineers. Sammy and I flew north in bitterly cold weather—-50 to -55°F on the ground most of the time. I again landed at Yellowknife, on December 29, and rendezvoused with Johnny Baker at the foot of his rocky hill. I told him to have his outfit ready to load quickly on my return from Great Bear Lake in a few days time. We had a number of short trips to do in the Great Bear Lake area, to Camsell River, Hottah Lake and Contact Lake, but finally we landed back at Yellowknife to pick up the Baker crew. It was January 4, 1935, and the temperature was about -55°F.

Sammy and I left the engine ticking over and trudged part way up the hill to help Johnny and his men carry their gear down to the aircraft. We worked as fast as we could because of the cold, but both Sammy and I froze one side of our faces before everything was loaded aboard. It was getting close to dusk and a light snow was beginning, so we took off as quickly as possible to try to get into Fort Resolution before full darkness had set in.

We were all jammed in the aircraft, half frozen. I, of course, was in my cockpit in front. The prospectors were sitting on and around their equipment in the middle, while Sammy was jammed in the rear of the cabin. During the flight to Fort Resolution, I felt the aircraft nose up sharply a few times. By using the stabilizer wheel alongside my seat I corrected each manoeuvre, but couldn't understand what was causing it. It turned out that one of Johnny Baker's men, sitting right behind my seat, was turning the stabilizer wheel from behind. Sammy could see what was happening but couldn't get forward to tell me. I finally caught the man in the act and stopped his antics. We groped into Fort Resolution and remained overnight. After a discussion with Johnny Baker, I learned that his man had been acting a bit queerly and was in need of some professional attention. Needless to say, we moved him to the rear of the cabin for the flight into Fort McMurray the next day.

From here on until the spring break-up, late in April at Fort McMurray, there was a steady routine of flights to the Great Bear Lake area and down the Mackenzie River, with the occasional day at home with Lin. Frank Kelly had been taken off CF-ATZ as air engineer and had been replaced by W.J. "Bill" Jacquot, a quiet-spoken man and a capable engineer.

13

Padres, Mails and Murder

Break-up was over at Fort McMurray by May 13, 1935. CF-ATZ, changed to floats, was launched and we commenced operations to the north again as far as the break-up would allow—Fort Chipewyan, Fort Fitzgerald, Fort Smith and Fort Resolution. Beyond that, the ice was just beginning to relinquish its grip. By June 3, we managed to get down the Mackenzie as far as Fort Norman.

A most important passenger flight was to begin on June 13. We were to carry the Father-General of the Oblate Order of the Roman Catholic Church and his staff on a tour of missions in part of northern Saskatchewan. The Father-General and an assistant had just come over from Rome for the trip. The company had tried to have an RC pilot carry out the operation, but it turned out that all the pilots in our area were Protestants. Those of us available matched to see who would go and as it turned out, the flight was to be mine. Bill Jacquot made sure that CF-ATZ was in good order and was clean and shiny.

The party consisted of the Father-General and his assistant, and Bishop Breynaut and an assistant. Breynaut was the Roman Catholic bishop for our northern area and was much liked by both the civilian population and our flying group. Bill Jacquot and I, as crew, made up the rest of the party.

This operation, lasting from June 13 to 16, took us to such places as Portage La Roche, Ile La Cross, Beauval, Petuanak and Buffalo River. At each place the missionaries turned out with their congregations for parades and special services. Bill and I were well looked after, taking our meals with the Father-General's party and always accommodated in the mission houses. The only discomforts were inflicted by hordes of mosquitos.

At Beauval we were met by some of the fathers in a large freighter canoe bedecked with flags in honour of the Father-General. In charge of the outboard engine at the rear was a big, strapping man, a Brother, in his black soutane. We were placed in the seats of honour in the canoe and then the Brother tried to start his outboard engine. He yanked the cord many times but it wouldn't go; then he started to swear, loud enough that all could hear. There was a frozen silence for a minute, but finally it became obvious that all of us, including the Father-General, were trying to restrain ourselves from bursting into laughter. Eventually the engine started and we headed for the mission.

On our return to Fort McMurray we were given the job of resupplying the posts at Wabiskaw, Trout Lake and Chipewyan Lake, southwest of the base, which kept us flying steadily until the 25th, when we were instructed to go back on the Great Bear Lake run for a trip or two.

On one of these trips we carried, along with a load of freight and mail, a British journalist, Irene Biss. She was anxious to go to the celebrated Great Bear Lake region to write a story about the hard rock miners. It was getting close to the time of the annual party there, the Miners' Picnic, which sometimes got a bit rough. I was dubious about taking her in at this time, but it had to be done. We made all the stops at the various points en route, and finally landed at Cameron Bay. There was no commercial accommodation at this time suitable for a lady staying overnight. Mrs. Reed of the Lakeshore Inn was there, but she lived in private quarters with her husband. There were one or two other ladies there by then also, but they too lived in cabins with their husbands.

Harry Reed and his wife curtained off a corner of the public living room-dining room with some sheets of canvas and placed an eiderdown sleeping bag on the floor. (We always slept on the floor in sleeping bags at Cameron Bay.) Walter Gilbert, our chief pilot, was also there on a flight, so he and I laid out our eiderdowns on the floor outside Miss Biss's compartment. In this way, we would be able to keep people from bothering her. During the long 24-hour daylight that night, we heard music coming from outside the building. It was one of the miners, very much inebriated, singing and playing a guitar outside her window. He kept it up for some time, and whenever he faltered, Walter and I would cheer him on, and off he would go again. The following day, we flew Miss Biss to Contact Lake to a quiet type of hard-rock mine, where she could get on with her writing in peace. A day later, I picked her up and she flew south with us to Fort McMurray. She apparently enjoyed her trip and published her writings successfully.

On July 1, we received word that Lesser Slave Lake and area (between Edmonton and Fort McMurray), and also the waterways of the Peace River, were in heavy flood. The railroad to the Peace River area from Edmonton was washed out in places, including some bridges. Since there was no suitable road to the area, badly needed supplies had to go in by air. A.M. "Matt" Berry and I, each with a Fairchild, were ordered to fly to Edmonton and begin emergency flights from there to the Peace River country.

We were in the air from three to eight hours a day carrying food, medical supplies, people and emergency mail to various points in the Peace River area—Grande Prairie, McLennan and the town of Peace River. It was hard flying in that a good part of our route was over comparatively dry country—a bit hard on the nerves when on floats. But our engines kept running and the weather was reasonable. By July 15 temporary rail service was restored, so we were ordered back to Fort McMurray.

Word reached us that two of our pilots, Herbert Hollick-Kenyon and J.H. "Red" Lymburner, were to go with the Lincoln Ellsworth expedition to attempt to fly across the south pole. Accompanying them as spare pilots were Bert Trerice and Pat Howard. We all wished them the very best of luck on their hazardous adventure.

My aircraft was undergoing maintenance and since Con Farrell was away, I was told to take his Junkers, CF-AMZ, with engineer Frank Hartley, on a short Barren Lands job. It was the re-supply of a trading post at Windy Lake again. This time, however, the total load was small, requiring one flight from Stony Rapids to Kasba Lake with barrels of fuel. Once the fuel cache was established, Frank Hartley and I did four round trips from Stony Rapids to Windy Lake, which completed the operation. We left Fort McMurray July 27 and returned the evening of the 31st.

Business was a bit slow at this time, allowing Walter Gilbert to get away for a short while. I was asked to fly his Junkers, CF-ARI, to Edmonton on August 10. I was very interested in Walter's Junkers as it had recently had a new engine installed. In place of the Pratt and Whitney Hornet, a P & W Wasp SC-1 had been installed. This was the first of the more heavily supercharged engines to appear in our area. With it we didn't open the throttle fully on take-off, and instead opened it only until a certain manifold pressure showed on a gauge. I was impressed with the performance of the re-engined Junkers. On the return journey I picked up Archie McMullen's Bellanca, CF-AIA, and flew it back to Fort McMurray.

About this time, my brother Dick was issued his first air engineer's licence. This meant a boost in salary and allowed him to start flying north as part of an air crew. He first flew as engineer with Rudy Huess and then later with Paul Davoud, who had recently joined Canadian Airways.

On August 17 I was back again on my Fairchild, CF-ATZ, with Bill Jacquot as engineer. We made a number of trips between Fort McMurray and Edmonton with special loads, and on the 29th we went north along the Mackenzie with mail as far as Fort Norman. On the return I was instructed to unload my southbound load at Fort Chipewyan and take on a special load to the new mineral area, Goldfields, on the north shore of Lake Athabasca, where a settlement had sprung up overnight. The mail carried on that flight, September 2, was the first official air mail between Chipewyan and Goldfields and return. I was then sent back into Great Bear Lake area with another load. A flight with Inspector Tom Reilly of the Post Office Department from Great Bear Lake to Coppermine on Coronation Gulf on the Arctic coast followed on September 4. It took us over quite forbidding looking country.

On our return we learned that Con Farrell and Frank Hartley in Junkers CF-AMZ were missing. They had been returning from Musk-Ox Lake in the northern Barren Lands after landing a party of trappers and their supplies. They were heading for Fort Reliance, at the northeast corner of Great Slave Lake, when they ran into an early fall blizzard, finally landing almost out of fuel on Frye Lake. Due to weather they were not sure of their location. They made use of a radio, which was an innovation in the company, a small battery-powered set with a Morse Code key. There were no aerials on the aircraft as the radio could only be used on the ground. A long aerial wire was thrown over tree branches as high as possible when the radio was to be used. Con made contact with Fort Rae but was unable to give his position.

Matt Berry was placed on the full-time job of searching for Con, and the rest of us were told to veer over into the general search area whenever we were nearby. After 11 days of searching, Matt finally spotted smoke rising from a fire which Con had set on a small island. Then he saw the Junkers on the shore. He pretended that he did not see the lost party, flew on by and disappeared from sight. Con, who was noted for his temper, apparently put on a good display, thinking that it was my Fairchild which had flown by and that Bill Jacquot and I were half asleep when we should have been looking around. After letting Con and Frank Hartley worry for a bit, Matt flew back around at low altitude and landed. Con and Frank were relieved to get back into circulation again.

From now until the latter part of October, when freeze-up started again, we flew steadily on our runs down the Mackenzie to the Great Bear Lake area, and also east from Fort Chipewyan along the north side of Lake Athabasca to Goldfields.

During this freeze-up period, Lin and I took a month's holiday. We motored from Edmonton to Oliver, in the Okanagan Valley of B.C., with Doctor and Mrs. Herbert Heal. Herbie Heal had been a fellow junior officer in the 20th Field Battery in the twenties, and later, as a dentist, did a lot of work in the far north, sometimes travelling in my aircraft. During one of our flights we had planned our joint holiday in Herb's new Hudson Terraplane car. On our drive back from the Okanagan, we went to Radium Hot Springs in the Windermere country to join Dot and Stan

McMillan of Mackenzie Air Services for the rest of our holidays. The motor trip was delightful despite the narrow gravel roads through the mountains. We returned to Edmonton, rested and refreshed, for another winter in the north.

On November 19 I was to test fly our one and only Lockheed Vega, CF-AAL, before taking it back to Fort McMurray. The Vega was a beautiful monoplane with a circular plywood fuselage, plywood wings and a Wright Whirlwind engine. We could only use it on skis or wheels, as it had not been fitted with floats. I flew it on skis around Edmonton for half an hour and found it a joy, smoother and a bit faster than the average aircraft of the day. Two days later, with Lin and a full freight load aboard, I flew the Vega to Fort McMurray in 2 hours and 20 minutes.

By November 24 my own aircraft, CF-ATZ, was ready to fly again, so Bill and I went back to a winter run to Great Bear Lake and vicinity. On December 1, on our way south, we arrived at Fort Chipewyan, where we were given some dreadful news. John Harms, a big, powerful trapper, had murdered his partner and then started a local reign of terror, sniping at some neighbouring trappers and their families. The crimes had occurred at Spring Point, on the northern shore of Lake Athabasca. Sergeant Pat Vernon, RCMP detachment commander at Fort Chipewyan, had started by dog team down the lake, accompanied by his Indian interpreter, and we were to give him all the aid we could.

We unloaded our aircraft and reloaded it with supplies for Goldfields, which was just east of the murder scene. John Harms' cabin was near Spring Point and back in the bush from the lake. We took off, flying east along the north shore of the lake. Nearing Spring Point I spotted the RCMP dog team still travelling east. I flew over it low, waggled my wings, then set course for the area of Harms' cabin. It was well back in the bush with only a narrow path leading from the lake to the clearing where it stood. As I circled, a figure, obviously Harms, came outside and waved a white rag. Assuming that he intended to surrender quietly, I flew on to Goldfields where we unloaded our freight. As it was getting dark we remained overnight. At dawn we took off again and headed back over Harms' cabin, but this time he came out with a rifle in the crook of his arm and without any sign of a white flag. Was this to be another shootout, *a la* Albert Johnson of Rat River fame?

I flew down the lake, landed on some ice near the RCMP dog team party and told Pat Vernon what I had seen on both flights over the cabin. We decided the man must still be dangerous. Leaving the interpreter with the dog team, Sergeant Vernon boarded the aircraft and we took off again for Harms' cabin. I circled low over the cabin and saw him in the clearing, still with his rifle. Then I found a fair place to land away from the point where Harms' narrow path joined the lake, for we didn't wish to line ourselves up with that path in case he decided to do some long-range shooting.

We left our engine running in order to keep it warm and also to be able to get off fast if we came under fire. Bill Jacquot was to sit in my seat and keep everything ready to go. Pat Vernon decided that he and I would enter the bush on either side of the pathway and work up through the trees, joining forces at the cabin clearing. We each carried a rifle. After we had left the aircraft Bill decided he did not want to be alone, so he followed along behind.

I can still recall my feelings as we started across the ice for the shore, and I am sure Pat Vernon can also. It was a case of waiting to be shot, out in the open, by a man concealed in the bush. We kept going without mishap, however, then separated to go through the bush on either side of the path. Once we entered the bush there was a terribly lonely feeling. We were now far enough apart that we could not even hear each other's movements in the deep, exhausting snow.

I finally arrived near the clearing, but I was so weary I could hardly get my breath from pushing through the snow. I decided to stop for a moment to catch my wind, leaned against a tree, holding my rifle out of the snow, and closed my eyes while I sucked my breath in. When I opened my eyes I turned rigid, for there, standing in front of me with his rifle in the crook of his arm, was John Harms, the killer. I was holding my rifle with its muzzle straight up in the air, an easy mark if he had decided to shoot.

What followed was anticlimax. He looked at me and said, "I guess you have come to get me." I answered, simply, "Yes." He handed his rifle to me and I told him to go back to the clearing. I followed, carrying both rifles. As we reached the clearing Pat was just coming out of the cabin. He snapped handcuffs on Harms.

We took Harms down to the aircraft and put him in the rear of the cabin. Pat and Bill sat with their backs to my seat so that Harms could not get at me while I was flying. We landed at the police camp where the interpreter was looking after the dog team. Pat assured us everything was under control, that he and his interpreter could handle Harms, so Bill and I took off, headed back to Fort Chipewyan, picked up my old freight load and flew back to Fort McMurray, arriving home on December 3.

Later Gen. Sir James MacBrien, then the Commissioner of the RCMP in Ottawa, sent a letter of commendation about my part in the episode to the owner of Canadian Airways, James A. Richardson. I received a copy with a covering letter from our general manager, Tommy Thompson. I prize both of these letters highly. John Harms, the trapper, was tried and sentenced to death for the murder of his partner. He was hanged at North Battleford, Saskatchewan, in May 1936.

Walter Gilbert, our chief pilot, told me that I had been selected to go to the Boeing School of Aeronautics at Oakland, California, to take a full course in instrument and night flying, including the latest in modern airline operations. It would take about three months to complete. On my return to Canada I was to organize an airline training school for our own Canadian Airways pilots, in anticipation of the company getting a trans-Canada air mail contract. I was so elated by these developments that I was almost walking on air.

I continued flying on the Great Bear Lake route throughout December. On the last trip we were southbound just before Christmas and had picked up our old friend Bishop Breynaut en route. We spent the night at Fort Smith, the Bishop's headquarters. When we had secured the aircraft for the night, he asked me what I had in mind as a Christmas present for Lin, knowing well that I hadn't had any time or opportunity for shopping. I said I wasn't sure yet. He took me to the mission and showed me some of the most beautiful Indian work I have every seen. From it he selected a pair of beautiful, rare, white caribou skin gloves, nicely embroidered, saying, "There is your present for Lin." I thanked him and later gave the gloves to her, a gift she always prized. During my absence Lin had arranged the sale of our bits of furniture and had almost completed preparations for leaving Fort McMurray.

14

Student and Schoolmaster

On January 2, 1936, after saying goodbye to everyone, including my brother Dick (now fully occupied flying as an air engineer), Lin and I flew to Edmonton where I reported to Punch Dickins. He signed my flying log books to date, ending that part of my bush flying career, and wished me luck in the new venture.

Lin and I climbed aboard a train from Edmonton to Vancouver, where we boarded a United Airlines Boeing 247 airliner for the south. The Boeing 247 was the most up-to-date airplane we had seen. I was intrigued by the way the pilot flew through heavy weather on instruments, as this was what I must now become proficient at.

We landed at Oakland in warm sunshine after leaving Fort McMurray a few days earlier, where it was -50°F. Awaiting us was George Myers, chief of flying for the Boeing School of Aeronautics. We found an apartment in nearby Alameda, purchased a second-hand Essex car and were then ready to go again.

The Boeing School consisted of a number of hangars at Oakland Airport containing many training aircraft and a variety of classrooms. We spent a given number of hours per day in class and the rest, weather permitting, in the air. Our studies covered airline meteorology, airline navigation and astro navigation, in addition to scheduling of flights and so on.

For the flying periods, we started our instrument flying in Boeing 203 biplanes, with Wright J-6 or Lycoming engines. As we became more proficient, we were changed over to a larger plane, the Boeing biplane type 40B-2 and 40C, with Pratt and Whitney Hornet or Wasp engines. These were similar to the aircraft used by Western Canada Airways on the old Canadian prairie night mail service in 1929-30.

There were many students on course. Some were starting their initial training; others, like myself, were taking special courses; and there were even a few officers of the Royal Air Force taking special instrument training.

I worked as hard as I could to make a success of the course, under the guidance of George Myers, Pop Gregg, the senior instructor, and Dick Field—a knowledgeable and helpful team. The most outstanding point of the Boeing instrument course was that we used banked turns. We flew, when on instruments, in just the same manner as when flying in the clear. In the instrument flying training I had taken at Camp Borden we used flat turns when flying blind, which I had always considered dangerous. As a result of this course, I instituted banked instrument turns in Canada on my return.

While at the Boeing School, I met H.T. "Slim" Lewis, then with United Airlines, which was associated with the Boeing enterprises, and P.G. Johnson. Later on, Johnson was to become the first vice president of Trans-Canada Air Lines, and Slim Lewis was to become its chief technical advisor—flying. We were to do a lot of work and flying together. Also at Boeing School were "Nat" and "Boo," sons of Bill Boeing, the owner of it all. They were a very fine pair of young men.

On weekends, when there was no flying, Lin and I would explore our part of California in the Essex. We had some pleasant drives through very pretty country, particularly around the Santa Clara valley.

The course ended for me on March 10, 1936, by which time I had completed 30 hours of instrument and radio range flying, in addition to night flying. The general manager of the school, T. Lee, Jr., sent a letter to the general manager of Canadian Airways Limited, telling him that I was now fully qualified in all respects regarding airline instrument flying and airline operations. A copy was given to me to keep which, of course, I still have.

We sold the Essex and flew back to Vancouver with United Air Lines, then took a train to Winnipeg. There I was to organize a training school for our pilots. Some changes had taken place while I was away. Punch Dickins had been transferred from Edmonton to Winnipeg to become general superintendent of all northern operations, and Wop May had taken Punch's place in Edmonton. Walter Gilbert had been transferred from the position of chief pilot at Fort McMurray to that of superintendent of the north Saskatchewan area, with headquarters at Prince Albert.

The company allotted me two Laird LC-B 200 biplanes, CF-AQY and CF-APY, to start our Canadian Airways Instrument Flying School. Engineer Al Dyne was responsible for the aircraft and engines, and two instrument specialists, Mac McMasters and Jack Hughes, looked after the specialized instruments in the aircraft. They removed existing instruments, gave them special calibrations for instrument training, then re-equipped and refitted both cockpits in line with Boeing School standards. A folding canvas hood was fitted over the front cockpit. A small radio receiver was carried for use on radio ranges (the airline radio navigation system) but there was no two-way radio aboard; also there was no generator on our

Wright J-5 nine-cylinder engine. This meant that when the aircraft battery ran low from using radio and lights, there was no way of recharging it in the air, which caused some difficulty at times. The school used the Canadian Airways metal hangar at Stevenson Field for its Winnipeg operations.

The Laird was a heavy, stubby aircraft for its size, and cruised at 110-120 mph. It had a tail skid at the rear which slowed it down a bit on landing, but no brakes of any kind. Its spinning characteristics were a bit frightening, particularly as Canadian regulations required each pilot qualifying in instrument flying to be able to spin his aircraft blind in both directions and recover. This was difficult and a bit hair-raising with the Laird. It would spin in one direction quite normally for three or four turns, then suddenly tighten its spin and go down like a corkscrew. In the opposite direction it would only spin about a turn and a half, then the spin would cease, and suddenly it would kick rather violently into a spin in the reverse direction. This caused some difficulty and some amusing incidents in the process of passing our pilots out in this manoeuvre. Our aircraft riggers tried hard to improve its spinning characteristics, but without success.

The instrument flying hood on the front cockpit proved awkward for myself as instructor in the rear cockpit. When the hood was pulled over the cockpit, it was like a canvas balloon, as its steel framework caused it to assume a roundish shape. I sat immediately behind so that my field of vision was limited and I couldn't see over or around it, but only to the sides. It was necessary to change course frequently to have a look ahead for reasons of safety.

Lin and I had found a boarding house in Winnipeg and were comfortably installed there. As a result of our experience in California, we purchased another second-hand Essex. Everything was in shape to get the school under way, including an information letter which had been sent to all Canadian Airways pilots advising those interested to volunteer for the course. They were to provide themselves with a special book about aeronautical meteorology at a cost of $3.00 and one about air navigation by Weems for $4.75.

The first course in Winnipeg started with five pilots: Paul Davoud, Stan Wagner, Ronald George, David Glen (later Regional Director, Department of Transport) and M.B. "Jock" Barclay. Training operations started on April 21, 1936, and from there on it was very busy indeed. Up to five hours per day were spent in the air, plus an hour or two in the classrooms where, with the aid of a blackboard, I taught radio range procedures, orientations, meteorology and navigation, including astro navigation. On certain nights I gave flying instruction, which involved the use of landing lights and parachute flares, which were in use at that time.

There was a difference of opinion between the Department of Transport and myself in the matter of banked turns versus flat turns when flying on instruments. Flat turns were still officially approved, but after some discussions on the subject, I was allowed to proceed with my banked turn instruction.

I had made an application for the new public transport licence, which was intended as an airline pilot's type of licence because instrument flying capabilities were required, and which was to supersede, when granted, our present commercial licence. I was very pleased when a registered letter arrived enclosing my brand new licence. All tests had been waived. The number was 24. Twenty-three licences had been given to Department of Transport inspectors, so mine was really the first commercial one. Included in it also was a renewal of my flying instructor's licence. The public transport was the licence that would be issued to our Canadian Airways pilots who graduated from my Instrument Flying School.

Of particular interest to Canadian Airways pilots at this time (May 1936) was the safe return of pilots Herbert Hollick-Kenyon, J.H. Lymburner, Bert Trerice and Pat Howard, after successful but very difficult operation flying across part of the Antarctic, including the south pole, thus completing the objective of the Lincoln Ellsworth Expedition. They were feted in Winnipeg before returning to flying duties. For their achievements, the Canadian government granted Hollick-Kenyon the rank of Honorary Air Commodore and J.H. Lymburner the rank of Honorary Group Captain in the RCAF.

The first Winnipeg instrument flying course was finished by July 22. All the pilots taking the course were tested and passed by Inspector T.M. "Jock" Shields of the Department of Transport, who did all our Winnipeg testing.

The next course was planned for Vancouver. There I could train some of our western division pilots without taking them too far from their duties. The ground school equipment was boxed and shipped to Vancouver. I was to fly CF-AQY to the Coast for the job. Lin wanted to visit her parents in Medicine Hat while I was busy on the West Coast, so I planned to fly her as far as Medicine Hat in the front cockpit.

On July 29, 1936, we took off from Stevenson Field for Regina, which required 3 hours and 30 minutes flying time. On the 30th, we landed in Medicine Hat from Regina after 3 hours and 40 minutes. Here I left Lin behind and flew to Lethbridge to see my parents. Weather interfered until August 2, when I took off for the flight through the Rocky Mountains. I climbed to 9500 feet above sea level, but could go no higher because of the weather. There were no radio ranges in the mountains, so I was forced to stay under the clouds. After following passes and valleys in fairly turbulent air, I landed on a small plateau used as an aerodrome, called Columbia Gardens, near Trail, B.C. It was a challenge to stop the

Laird, which had no brakes, before rolling over the edge of the plateau! Flying time on this leg was 3 hours. Next day I was limited by weather to a maximum of 8000 feet above sea level, but finally groped my way to a landing in Vancouver after 3 hours and 15 minutes. I immediately got busy organizing the Vancouver school.

A small group of Canadian Airways pilots was selected for the course, including E.P.H. "Billy" Wells, F. Maurice McGregor and Herbert Hollick-Kenyon. "Ken" was now involved in bringing up two twin-engined Lockheed Electra ten-passenger airliners from Burbank, California, to Vancouver. Between flights, he took the instrument flying course in Vancouver. In September he was to undertake a series of flights across the mountains for the Department of Transport in preparation for scheduled air operations. He would use one of the new Electras which he and I would bring up from Burbank.

On August 5 I started daily (weather permitting) training flights with the Laird. Hollick-Kenyon arrived with the first of our Electras, CF-AZY. He checked out Billy Wells and Maurice McGregor on the aircraft, after which it flew on our regular scheduled service between Vancouver and Seattle. In the second week of August, I managed to get checked out on the Electra at the Vancouver end of its run, which was in addition to flying the Laird part of each day.

Ken was now ready to go back to Burbank for the second Electra, and I was instructed to drop the training temporarily and go with him as second pilot. Al Dyne was to be our engineer.

We arrived in Burbank to find that our aircraft, CF-BAF, wasn't quite ready. The Lockheed Aircraft Company had arranged fine accommodation for us during our stopover, toured us around the countryside in a chauffeur-driven car, and wined and dined us each evening. On August 20 we flew with a Lockheed test pilot to Las Vegas, there to officially take delivery of the second Electra, which was done to avoid a California state tax. The next day, we flew from Las Vegas to Seattle, climbing to 14,000 feet; the flight took 5 hours and 50 minutes. The following morning we flew from Seattle to Vancouver in 50 minutes.

For the next three days I flew with Billy Wells and Maurice McGregor on the Seattle run, where we did instrument and radio range training during the flights. On our return to Vancouver, I would start up the Laird and we would do more training with it.

On August 27 I flew with Ken to Lethbridge in Electra CF-BAF. We flew at 13,500 feet and completed the trip in 2 hours and 50 minutes, record time.

Back in Vancouver again, our training and flying was divided between Electra, CF-AZY, and the Laird. Ken had completed his instrument training and had now taken CF-BAF east into the mountains on his

Department of Transport testing operation, with, I believe, S/L J.H. Tudhope of the DOT. I carried out training in Vancouver until September 12, when I flew the Laird to Seattle. Here we were able to accomplish considerable work using the Seattle radio range, and here, too, we did much night flying. I returned to Vancouver September 22, where the final DOT tests were done on my students by Carter Guest, the local DOT inspector. When all had successfully passed, it was time for me to move on again.

On September 29 I loaded the front cockpit of the Laird with my gear and took off eastbound late in the day. As I was nearing Princeton, I found myself getting dizzy. It had the earmarks of my old dizzy spell of 1932, and was somewhat reminiscent of the incident while flying into Island Lake in October 1934. I put my head down as much as I could once or twice while still trying to keep the aircraft level. Finally my head seemed to clear and all was well again, but I could not understand what was causing this.

I got as far as Princeton before dark and remained overnight. The following day, with a full load of fuel, I took off and made Lethbridge non-stop in 5 hours and 25 minutes. The latter half of the trip was through the mountain valleys at low altitude because of bad weather; a wrong turn in certain places could have been disastrous. In flying through the Crows Nest Pass at low level and in poor visibility, I was startled to have my engine cut out momentarily a couple of times. It was checked over in Lethbridge that night, but everything seemed normal. (I reported this later on in Winnipeg, but the trouble could not be located. Very much later on, I was told by the engineers that the J-5 engine's cam ring had worn a bit loose and when the carburetor iced up slightly, as it was doing in the mountains, a slight backfire would throw the timing out momentarily and cause it to stop for a second or two.)

From Lethbridge I flew to Medicine Hat to pick up Lin for the journey east. The weather was getting cold now, so Lin was dressed in warm heavy clothes, ski pants, Hudson's Bay coat, and heavy socks and boots. Her front cockpit was even colder than my cockpit. We took off from Medicine Hat, again with full fuel, and flew direct to Winnipeg, with a bit of a tail wind. We were both stiff with the cold when we crawled out of our cockpits at Stevenson Field. It was October 2, 1936.

The next assignment, after having work done on my aircraft, was to fly to Montreal and Charlottetown, there to train two more groups of Canadian Airways pilots. My instrument flying school was becoming a veritable flying circus.

15

Lunch-time on the Range

I shipped my box of ground school equipment to Charlottetown and the Laird was declared ready for operations on October 8. Lin and I packed our gear and readied our heavy flying clothes for departure. Aerodrome facilities were then sketchy across northern Ontario, so, following custom, I took the American route around the southern end of Lake Michigan with re-entry into Canada at Windsor. The facilities for aircraft service, particularly in the cold weather, were better south of the border. Within the next year or so, however, this would change radically.

We took off on October 9, Lin and our baggage jammed in the front cockpit, and both of us wearing our bush flying gear to be as warm as possible. Our route took us, over four days, to Pembina and Fargo, North Dakota; St. Paul, Minnesota; Madison, Wisconsin; Battle Creek, Michigan; and finally to Dearborn Airport, then controlled by the Ford Motor Company, just outside Detroit. By flying on the United States airway I was able to use my radio range receiver and practice and maintain my personal flying standard. We stayed at Dearborn the night of the 11th, planning to leave the next day. However, when we were ready to depart, I found that I couldn't: it was Sunday, and the Ford people would not permit any operations to take place on the Sabbath.

Monday, the 12th, we flew to Toronto, where our general manager, G.A. Thompson, told me to speed up my training program if possible, as things were developing in regard to a possible trans-continental service. I replied that I would do my best without reducing the high standards of the courses. On the 13th we flew to Montreal.

On the 14th we landed at Quebec City for fuel, then flew to Moncton —total time for the day, 5 hours and 10 minutes. We were so cold that there was no question of going any further that night. On the 15th we flew to Charlottetown, where the company's chief pilot, Walter Fowler, was waiting to meet us. The Charlottetown field was not very large and had a grass surface from which Canadian Airways operated several services, including to the Magdalen Islands. The chief aircraft here was the de Havilland Dragon, a twin-engined biplane. Lin and I took a comfortable room in the main hotel, where, as old prairie people, we took the opportunity to eat Malpeque oysters and other sea foods whenever we could. In addition, there was no shortage of good Demerara rum.

Early the following morning, October 16, we started flying training operations. The pilots to be trained were Walter Fowler, H.S. "Jonesy" Jones and Bert Trerice. Bert had been a company engineer and now had become a good pilot. There was no radio range available, so we had to do simulated exercises.

Good weather permitted flying every day, plus a fair amount of night flying, so Walter Fowler and Jonesy soon completed their training. I wanted to get back to Montreal, where I had a larger group to train, and decided to take Bert Trerice with me to finish his training en route. Lin and our baggage went to Montreal by train.

I took off from Charlottetown on October 30 with Bert Trerice in the front cockpit. My plan was to arrive at a small airport in Maine just before dark, refuel and go on to Montreal that evening, giving Bert the night training. In my cockpit were two controls for the big parachute flares which could be used in emergencies at night.

After take-off, I had Bert Trerice close his instrument flying hood and start work. He flew on instruments, and I merely gave him course changes, as there were no navigation facilities east of Montreal. There was an old loop-type radio range in Montreal, which I had planned to use when within range later that night.

As we flew on, the weather slowly deteriorated. I realized that we should have taken off earlier as, when we got into the mountainous areas of Maine, it was starting to get dark. I had never seen the small airport at Brownville, Maine, before but figured that I could find it easily enough. By the time we had arrived in the vicinity of Brownville, it was dark and was beginning to snow. I had had Bert open his instrument hood a short time earlier, and now he was to get some real night instrument training.

It was impossible to see the little field below us in the darkness and snow. Moreover, it was essential not to stray too far to either side of our course because of the high hills. I told Bert I would show him how to find the airport and land under these circumstances. The procedure was to drop a parachute flare at an altitude of 4000-5000 feet to show us the general area and locate the airport, then drop another flare at 1500 feet or thereabouts, right over the airport. We would then quickly circle while the flare was still descending, and complete our landing with our landing lights. Bert said he would watch the procedure carefully.

When I thought I was over the right spot I pulled the first flare release, telling Bert what I had done. We waited, but nothing happened. I informed Bert that this was a case of flare failure, but that the second one would surely work. I lost altitude to about 1500 feet to be in position for the next one, hoping that I could keep away from the hills around me. I pulled the second release, but there was nothing but darkness and light snow. Bert asked me what we should do. We had to find the airport, which had no lighting, with the landing lights, using them to sweep the area at very low altitude. Luckily I saw what looked like the field, circled and landed fast, as the landing lights would not have lasted long without a battery charge, there being no generator on the engine. We taxied to the edge of the field where there was a car with its headlights on. We got out of our aircraft, rigid with cold after five hours in the air.

I crawled under the fuselage to look at the flare chutes and there, large as life, were the flares, securely pinned in position. Unfortunately, in the haste of departure, I had forgotten to remove the safety pins, which prevented the flares being dropped inadvertently ... and prevented me from using them when I wanted them. I never forgot that lesson. From there on I never embarked on a night flight without ensuring that my flares were usable. It was also a good lesson for Bert.

The next morning, October 31, there was low cloud, snow, and the possibility of ice. Ice was a particularly bad hazard on a biplane, with double leading edges on the wings, as well as wires and struts to ice up. There were no de-icers on the wings or propeller, nor any heat control to keep ice from forming in the carburettor. I figured that it was a good flight for Bert Trerice's instrument experience because we could do it without the instrument hood. Moreover, since we were going to St. Hubert, we could go in and approach on the radio range to the St. Hubert Field. It looked reasonable, so we departed from Brownville and climbed up into the overcast, setting course for Montreal. The ceiling at Montreal was reported to be around 1000 feet, so everything looked good.

We managed to hold our altitude and course although we were bothered intermittently by ice. Finally, as we got close to Montreal, we began to hear the St. Hubert radio range on our receiver. We were nicely located on the range and about 45 minutes out of Montreal when the range signals stopped and the operator's voice came on. He said it was 12 o'clock noon, and that they were shutting the range down while they went to lunch, and that they would be back by one o'clock. "Goodbye," said the voice, then everything went dead. I practically exploded with frustrated rage. With no two-way radio, I could not call to tell them that we were coming in on the range and needed it badly. Bert, who, I think, was beginning to wonder what sort of a business this new airline operation was, asked what we were to do now. I replied that we would hold our course and altitude for a while, and then slowly decrease altitude straight ahead so that we would be sure to be below 1000 feet before we got near Montreal. There was some danger of hitting one of the isolated hills in that area, but we had to chance that. We finally broke through the overcast just below 1000 feet, found St. Hubert and landed.

I reported the matter of the radio range and found that the shutting down procedure was done every day, as in those days no one was really using the range. After our experience, however, the range was operated through all lunch hours. To this day, I often have a good laugh with retired senior DOT officials over the matter of the range shutdown. Bert Trerice, however, had gained a lot of valuable instrument and night flying experience in a couple of days.

I met Lin in Montreal, then reported to our headquarters, where I met J.P.R. "Romeo" Vachon, the area superintendent. Romeo was a top Canadian Airways bush pilot. He had made all arrangements for my instrument school. The pilots who were to take the course were already in Montreal and, in fact, Romeo himself was going to take it too. He had arranged for us to stay over in Longueiul, not far from St. Hubert airport, in a large boarding house, with room for our whole instrument class, with the exception of Romeo, who lived in Montreal.

Romeo Vachon was to win the Trans-Canada McKee Trophy, Canada's highest award for civil aviation, in 1937. Later, he and I were both to take our places at the McKee table each year. It was always good to see him, until he finally passed away.

Our boarding house was owned by Mr. and Mrs. Crevier, whom we fondly called Pa and Ma. We were well looked after. Each pilot had his own room and three of us had our wives with us, namely Mary Bone, Jessie Lymburner and Lin.

The pilots on the Montreal course included Romeo Vachon, Dave Imrie, J.H. "Red" Lymburner, V.J. "Shorty" Hatton and F.W. "Freddie" Bone. Flying operations started from St. Hubert on November 4, 1936. We flew every hour that weather and time for ground school would permit. It was useful having a radio range near the airport, as we could carry out real instrument approaches without having to simulate them. Every second or third night, we did night flying. With the aid of Inspectors George Wakeman and Stuart Graham of the DOT we designed a radio range instrument procedure card for use at St. Hubert, the first made up in eastern Canada.

We flew very hard during November, December and the early part of January 1937, until everyone was qualified and tested. Stuart Graham, who had been bush flying as early as 1919, was the area air regulations inspector, so of course he tested my pilots. I can still recall talking to

Stuart just before he started to check the boys on their blind spins. I told him of the most peculiar manner in which the Laird would spin. Stuart looked at me and said, "Surely it can't be as bad as that." When he landed after carrying out the spinning manoeuvre, he said, "I just couldn't believe it, but it surely is so."

My work in Montreal was finished, but before I left for Winnipeg Romeo Vachon asked me if I would ferry a Fox Moth, CF-APG, a small biplane with a closed cabin for two passengers and an open cockpit in rear for the pilot, to de Havilland in Toronto for overhaul. This I did on January 17, 1937, returning by train to Montreal.

Lin had our baggage ready to go again, so I put her on the train to Winnipeg, as it was much too cold for her in the open cockpit in January. I took off in the Laird on January 23, 1937, and flew non-stop to Detroit, rigid with cold all the way. The next day, I flew to Battle Creek, then on to Chicago, where I stopped for the night, again a very cold flight. On the 26th, I flew to St. Paul, then on to Fargo in 5 hours and 40 minutes, which was all the cold I could stand. On the 27th, I landed again in Winnipeg, still in bitterly cold weather.

The following day, I attended a meeting at Canadian Airways headquarters with G.A. Thompson and Punch Dickins. It was decided to forgo further instrument training until spring, as it was really much too cold to sit under a hood in the front cockpit and try to concentrate on instrument flying. In any case, the Laird needed a major overhaul.

Herbert Hollick-Kenyon had returned to Winnipeg with Electra, CF-BAF, and it was decided that we would try an experimental scheduled airline service between Winnipeg and Red Lake, Ontario, using the Electra. A pair of fixed streamlined metal skis were fitted so that we could land on the ice at Red Lake, where, of course, no aerodrome existed. A small radio marker beacon was to be installed there and we had our old loop-type radio range at Winnipeg. We were to fly a daily service each way, with Hollick-Kenyon as captain and myself as the second pilot.

The Electra was ready for test on its new skis, so on February 15, 1937, Ken and I did a 55 minute test flight and found everything satisfactory. On February 17 we did our first round trip to Red Lake. The passengers were enraptured with the airplane, as it was undoubtedly the most luxurious most of them had ever ridden in.

We flew on a proper airline schedule, with flight plans, frequently flying on instruments and making our approaches right down to airport minimums at Winnipeg, with somewhat higher minimums at Red Lake. Our radio marker beacon was a big help at that point. Ken and I shuttled back and forth daily on this operation until March 29, when the job was finished.

It was now getting warm enough to use the newly overhauled Laird instrument aircraft, so a special training course was laid on to cover senior company officials, who were anxious to get a chance to secure their instrument training. This course ran from March 31 to May 10, 1937, and covered the usual instrument flying, night flying and unusual manoeuvres, including blind spins. The officials on course, all senior pilots, were G.A. Thompson, the general manager, Punch Dickins, the manager of northern operations, and V.H. "Val" Patriarche, the sales manager.

During this course, I was giving a lecture on radio navigation to a group of executives and university officials when suddenly, about halfway through the program, I became dizzy and collapsed. I came around quickly and finished my lecture, but I was mystified by what had happened. Again I did not report it to our medical men, as I was afraid they would stop my flying.

Early in June I was sent to Sioux Lookout, to help out at that base. The superintendent was A.N. "Westy" Westergaard. Pilots operating from there included Art Schade and Stewart McRorie. Our traffic agent was Fred Lundy, who had been with us at Fort McMurray for so long. I flew a trip or two in Junkers, CF-AQV and CF-ASN, also one with Stewart McRorie in the big Junkers "flying boxcar," powered by a Rolls-Royce engine. Finally I was given Bellanca Pacemaker, CF-BFC.

This Bellanca had one idiosyncrasy. Whenever I throttled back before landing, long flames would pour from its exhaust. No one seemed to know what the trouble was. It didn't do any harm, and the engine worked well, so I just carried on. However, we dubbed it the "flaming coffin." I flew loads out of Sioux Lookout to places like Red Lake, Gold Pines, McKenzie Island, Forestry, Pickle Lake and Argosy. We flew hard and long until most of the backlog was cleared, then on July 23, I was told to deliver my Bellanca to Vancouver.

With Al Dyne as engineer, I took off for Berens River on Lake Winnipeg, then on to The Pas and Prince Albert, where we stayed overnight. Our day's flying time was 7 hours and 35 minutes. The following day we refuelled at Cooking Lake, but were forced down by weather at Chip Lake, west of Edmonton. Later in the day, we took off again and flew through the Yellowhead Pass in the Rockies. We took a beating from turbulence and the weather closed in again, but we managed to land on the lake at Red Pass Junction and tied up for the night. The following morning, we took off early and after another rough ride landed and refuelled at Kamloops, and thence on into Vancouver. Here our Bellanca was turned over to the Western Division of Canadian Air ways headed at that time by Major D.R. "Don" MacLaren, the famous First World War ace.

Al Dyne and I were to return to Winnipeg by air the next day, which meant flying to Seattle aboard our own Canadian Airways Electra, CF-AZY, then from Seattle eastward via Northwest Airlines. We spent the night before our departure in a Vancouver hotel, where we had a room. That evening, some of our western pilot and engineer friends gathered there and we settled down to some steady drinking. We had a good get-together with the boys, but by the time they had departed Al and I were very much under the influence of alcohol, so we retired to our beds.

I woke up a bit later and through an alcoholic haze saw smoke rising from Al's bed. Apparently he had gone to bed with a cigarette still alight. I called but couldn't wake him, so staggered out of bed to douse the fire. In my inebriated state I tried to do it by pouring whiskey from an opened bottle on the night table. The fire then flared up and Al got out of the bed fast, and we managed to put the fire out, with water. In the morning we reported the matter to the hotel clerk, who said that we would have to buy a brand new mattress.

We were sure the hotel would merely get the damaged one repaired, but would charge us for a new one. The clerk was adamant about it, so we paid him for a new mattress. As we were checking out, we were still angry so decided that since we had bought a brand new mattress, the old one definitely belonged to us. We bundled it into a roll, tied it up and took it down into the lobby with our luggage. The hotel manager tried to stop us parading the roll of mattress across the lobby, but we insisted. We had a tough time getting it through the swinging doors, but we made it. Guests were standing around watching the performance and the hotel staff, of course, was shaken beyond belief.

Stuffing our gear and the mattress into a taxi, we left for Vancouver Airport. En route we passed an open area, so we stopped the taxi, deposited the mattress in a field and drove on, feeling better about the whole thing. We had a good trip back to Winnipeg, as passengers.

After a few days at home with Lin, I was sent to Prince Albert to help out in northern Saskatchewan with my old friend Walter Gilbert. By August 4, 1937, I was back in full bush operations on another Pacemaker, CF-BFD, and later a Fairchild 71, CF-AOP, flying to such places as Ile La Crosse, Goldfields, Stony Rapids, Lac La Ronge, Dore Lake and Beauval. While operating in this area, I met a young man named R.W. McNair who was a wireless operator for the Saskatchewan government. Buck McNair was to become one of our top Second World War aces. After the war, this courageous man made the RCAF his career and rose to the rank of group captain.

On August 17 I received a telephone call from Philip G. Johnson in Ottawa. The federal government had decided to form a new, government-sponsored airline to fly the trans-Canada route. All the work and training that we had done to prepare Canadian Airways for the contract was lost. The new company, to be called Trans-Canada Air Lines, was to be headed by Johnson as its first vice-president. He was well respected as an experienced airline executive by C.D. Howe, the Transport Minister, who had asked him to take on the task of building the new company. I had met Johnson while I was involved in my airline training at the Boeing School of Aeronautics, hence the telephone call from him.

He asked if I would like to join the new company as its senior pilot. I said I would have to discuss the offer first with my chief, G.A. Thompson. I did not wish to leave Canadian Airways without discussing the matter with them as they had been very good to me. He agreed and said that he would be calling Thompson himself. In a matter of an hour or so, I spoke to Thompson on the phone. He told me to drop everything and return immediately to Winnipeg.

In Tommy Thompson's office I was told that if I wished to carry on my airline work, it could only be done with the new company, as Canadian Airways was going to concentrate on bush flying again. He generously offered that if I didn't like the new company, I could always rejoin Canadian Airways, with my old seniority. I phoned Johnson from Winnipeg, accepting the job.

I was hired as of then, August 20, 1937, and there being no other pilots or engineers in the new Trans-Canada Air Lines as yet, I was despatched immediately to Vancouver to join a DOT party as the TCA representative. This party was setting out to determine the proper location of radio range facilities from Vancouver to Winnipeg and from Lethbridge north to Edmonton. I was to ensure that these ranges were placed in satisfactory locations for Trans-Canada Air Lines operations.

16

Building an Airline

I left Winnipeg the next day by train for Vancouver, where I joined up with the DOT party, which consisted of inspector of airways J.R. "Doc" Robertson, radio expert A.K. Bayley, and one DOT automobile. We were to travel mostly by car, but where necessary would use the DOT Lockheed 12, CF-CCT, piloted by S/L J.H. Tudhope and inspector Jack Hunter.

We drove to Princeton, where we took bearings and found a location for the radio range station, then on to Oliver, where we were joined by Tudhope and Hunter in the Lockheed. We planned the radio range location to be in the valley three miles south of the Oliver emergency landing strip (this plan was cancelled later on). August 28 Jack Hunter and I took off in CF-CCT to check the proposed location from the air. Because I was the one and only Trans-Canada Air Lines pilot, this meant the 30-minute flight constituted the first ever made by a Trans-Canada Air Lines pilot.

A few days later, on September 1, TCA took over the Vancouver-Seattle run from Canadian Airways, including two Lockheed Electras, CF-AZY and CF-BAF, and a Stearman biplane which had been used for dusting pea crops from the air. Pilots E.P.H. "Billy" Wells and F.M. McGregor were also taken on strength on September 1, as were engineers Sammy Reid and George Roper. The Vancouver-Seattle flight, operated on September 1 by Wells and McGregor in CF-AZY, was the first flight of an aircraft belonging to Trans-Canada Air Lines and the second by TCA pilots.

We drove to Grand Forks, B.C., located a good range site, and finished our bearings by September 2. A similar pattern was carried out at Creston, B.C. Here Tudhope and CF-CCT joined us again (the Creston radio range plan was also cancelled later). A suitable site was found at Cranbrook, B.C. Next we flew to Edmonton where we located a good site, finishing that job on September 8. A radio range station was not then considered for Calgary, so we passed it up.

During our stop-over in Edmonton, we had all gone shopping for some presents to take home to our wives, and among other things we purchased a few small vials of good perfume. In the hotel later, Tuddy and I decided to play a joke on Doc. We sprinkled some perfume on the lapel of his coat, which we knew would be hard to remove. He vowed to get even with us, and the next morning, as we were leaving for the airport, he offered me two or three Chiclet candies. I chewed them and later, in the air, found out that they were Feen-a-Mint—a strong laxative. Whenever I left the cockpit to rush back to the washroom, Doc would throw the aircraft around, making it difficult for me to hang on in the tail end. There was no question but that he had evened the score very nicely.

At Lethbridge Tuddy and CF-CCT left us. Our car had been brought there so we drove across the prairies, establishing radio range sites at Lethbridge, Swift Current, Regina, Rivers and Winnipeg. Later, one or two were added or shut down. The radio range job was finished by September 20 in Winnipeg. Now the DOT could begin installing the equipment.

On my return to Winnipeg I found that we had taken over Canadian Airways' large metal hangar and, of course, Electra CF-BAF, which Hollick-Kenyon and I had flown on the Red Lake run earlier. I found also that our first Winnipeg-based air engineer was there, grooming the Lockheed. He was Charles Grogan, one of Lockheed's factory engineers. He had decided to join Trans-Canada, and we certainly could use him. Unfortunately, he returned to the United States all too soon.

At this time it, it was difficult to know what was happening, as we had no administrative staff that I was aware of. By September 25 Charlie Grogan told me that CF-BAF was ready for a test flight. There were no other TCA pilots in Winnipeg yet, so I detailed Grogan to act as my second pilot. My last flight in CF-BAF had taken place the previous March, and other than the bit of flying I had done in CF-CCT, I was very rusty flying the airliner, having become more or less a bush pilot again from March on.

Between Charlie and me, we got the Electra into the air, flew it around and completed our testing. When I came in to land, the bush operations caught up with me. We had no flaps on our bush aircraft at that time so in landing I forgot to put the Electra's flaps down, and as well, my approach speed was too high. I put the aircraft on the ground successfully but found we were travelling very fast across the grass. The boundary fence loomed ahead, but by a judicious turn and hard use of brakes, we came to a standstill. It was then that I realized I hadn't used the flaps. Charlie made a few adjustments to the aircraft and we flew it again on September 27, this time properly.

The new TCA executive group started to arrive. They were D.B. Colyer as chief technical advisor; H.T. "Slim" Lewis as technical advisor, flying; O.T. Larson as technical advisor, meteorology and despatch; H.O. West as technical advisor, overhaul and maintenance, all of United Air Lines;

and W.A. Straith of Northwest Airlines as chief instructor. Shortly after came S.S. Stevens as technical advisor, radio communications. They were all fine, experienced people.

More Electras were ordered for quick delivery, and construction of a new hangar and headquarters office building was begun. Plans were made for training pilots, engineers and despatchers. We also started to hire air crews, many of them Canadian Airways pilots whom I had trained on the Laird.

It was now clear that I was going to be based in Winnipeg for some time, so Lin and I moved out of our flat to a rented house on Borebank Street and started to buy some furniture—an endlessly recurring theme.

Word was received that the first of the new Lockheed 10A Electras was due to arrive soon in Vancouver from Burbank. Most of our executive staff would go to Vancouver by train and fly back in the new plane. This would give them a chance to see what our new western route looked like. D.B. Colyer, H.O. West, H.T. Lewis, O.T. Larson, S.S. Stevens, W.A. Straith and I climbed aboard the train at Winnipeg for Vancouver. En route we worked from morning till night. Slim Lewis, Bill Straith and I completely wrote the first Trans-Canada pilot's manual for the Lockheed 10A, which we used for our first flying operations.

We worked hard but also had fun. Whenever the train began to slow down for a station, D.B. Colyer would start to sing

Passengers will please refrain
From flushing toilets while the train
Is standing in the station . . . or a stop.
We encourage constipation
While the train is in the station;
After that we let the matter drop.

He put the words to a catchy tune. Eventually we all did it, so that automatically as a station approached we would stop work and sing together our little refrain. It caused some amusement on the train.

In Vancouver we found that our new aircraft was CF-TCB. Lockheed CF-TCA was there also, but we decided to leave it on the coast for a while as a back-up for CF-AZY.

On October 10 we headed east in CF-TCB. Slim Lewis, Bill Straith and I alternated at the controls. We passed up Lethbridge due to weather, and landed at Medicine Hat after 4 hours and 10 minutes. On the 11th we flew to Regina, where we spent the night and also selected the site of the new hangar we planned to build there. On the 12th we arrived back in Winnipeg.

Slim Lewis, Bill Straith and I set up our training school in the old Canadian Airways metal hangar. We had two Lockheed 10As to start with, our old CF-BAF and our new CF-TCB. Slim operated as the chief of flying and part-time instructor while Bill and I started full-time instruction in instruments, radio range and night flying. We were hiring the best pilots we could get our hands on, some of whom were from Canadian Airways and were instrument trained. Others came from small companies and others were Canadians who had been flying in the United Kingdom.

Among the first pilots to join, aside from Slim Lewis, Bill Straith, Billy Wells, Maurice McGregor and myself, were Walter Fowler, Ron George, Bruce Middleton, David Glen, Bob Smith, Lindy Rood, Jock Barclay, Howard Sandgathe, Marlowe Kennedy, Herb Seagrim, George Lothian, Frank Young, Bill Coulson, Barney Rawson, Harry Umphrey, Jackie Jones, Jack Lewis, Ray Goodwin, Jack Wright, Art Rankin, Dave Imrie, J.R. Bowker and T.F.K. Edmison.

Concerned with despatching, under the control of O.T. "Ted" Larson, were Ted Stull and Frank Barager, both of them old Canadian Airways pilots. Frank Barager became our first despatcher at Stevenson Field by recording the arrivals and take-offs of our training aircraft. Our radio equipment was not yet set up in the hangar, so Frank installed an old radio receiver which covered our frequencies, listened to our reports and wrote them down, but couldn't communicate with us at all. S.S. "Steve" Stevens soon made the despatching staff effective as our new radio equipment started to arrive and operators were hired. Early in November we got another Lockheed 10A, CF-TCC, which I flew from Vancouver to Winnipeg.

Air engineers were being hired to staff the Winnipeg base, plus the field detachments at Regina, Lethbridge, Edmonton and Vancouver. My brother, Dick, was one of the engineers who joined in November 1937. He left Canadian Airways and came down from Fort McMurray to Winnipeg. Here he was taken on strength and checked out, then sent to Lethbridge as a base engineer.

We were all fitted with new blue uniforms, single-breasted with a belt. Our caps had white tops. The captains wore two gold rings on their sleeves and the co-pilots wore one. My salary was increased from the pay of a captain, $400 a month, to a special rate of $750 due to my being a senior instructor. The new pay rate was beyond anything I might have imagined and had a lot of buying power in 1937.

More pilots were appearing on strength, including Pat Howard, Bud Newcombe, Rene Giguère, Don Stevens, Gordon Haslett, Don Galloway, Don Brady and Jack Herald. We slaved through December 1937 to February 1938, flying each day and carrying at least two students on each flight. We covered instrument flying, take-offs and landings. The instrument flying included all manoeuvres up to and including stalls and unusual positions. On most nights we checked out our students on night flying, including dropping of parachute flares (after ensuring that the safety pins had been removed!).

By the time February 1938 arrived, Slim Lewis, Bill Straith and I were so tired we could hardly think. We were meeting our training schedules, however, having begun limited scheduled training flights to Regina and, by the end of February, to Lethbridge and Edmonton. These were day-time flights only, which sometimes carried ordinary mail. An arrangement had been made with the Post Office and the railroads that we would pick up a load of mail at Winnipeg, Regina or Lethbridge and transport it to the next point by air. Sometimes this speeded up the mail and sometimes, if the weather became unflyable, it actually slowed it down. No passengers other than company personnel were carried.

From early January to late in February 1938, in addition to our own pilot training program, we undertook to give the full course to A.D. "Dan" McLean, the Controller of Civil Aviation, and one or two of his executives. We were pleased to do this, as Dan and his men were able to help us with their thorough understanding of our work and problems. Dan graduated on February 17 and was given a special diploma, partly humorous and partly serious, but signed by all of us who were there at the time.

On February 27 Slim Lewis and I carried D.B. Colyer, Ted Larson and Steve Stevens on a flight to Kapuskasing as the first venture into our coming Eastern Division. We looked at the emergency fields en route but did not land on them. On our return to Winnipeg we started planning for our runs east.

Bill Straith left around this time to return to the United States (he was killed in a flying accident many years later). As a result Slim Lewis and I had to carry out the full instruction schedule, which was pretty rugged.

As well as our own pilots, we were now checking out some of the DOT inspectors on instruments as a result of A.D. McLean's enthusiasm for airline-type training. On March 20 I spent part of the day checking the inspectors out in the DOT's Lockheed 12, CF-CCT. The final stage of this was that inspector Jack Hunter and I were to try to spin the Lockheed. After getting plenty of altitude, I pulled it up and throttled its engines back. It kicked into a spin quite violently and gave us a bit of a ride down, but it came out again without too much trouble. We were both satisfied that one spin was enough, so went in and landed.

On April 1, 1938, we felt we were ready to start some limited scheduled night flying operations. We filled CF-BAF with pilots and Slim and I went along as check pilots. We flew to Regina and returned to Winnipeg. For a few nights I rode as check pilot, each night with a different crew, as far as Regina and back. The Regina flights continued to the end of April, then were extended to Lethbridge and later to Edmonton.

17

Mastering the Monster Lockheeds

Early in May 1938 we got word that the first two of our Lockheed 14Hs, the latest in airliners, were ready at Burbank. The 14s were bigger and faster than the 10As and carried 12 passengers in comfort, plus a crew of three, including a stewardess. A stewardess training program was to begin shortly.

Slim Lewis and I would pilot our first pair of 14s up from Burbank. Slim took Maurice McGregor as second pilot and I took Lindy Rood. We were well looked after by the Lockheed people in Burbank and accompanied their test pilots on a flight or two. The test pilots were unhappy about the violent stall characteristics of the new aircraft and its tendency to bounce badly unless landed correctly, which dismayed us a bit. We also found that we were not going to get a dual check-out on the 14 until we landed at Reno, where the aircraft were to be turned over to us officially.

We spent two days in Reno trying to fly the 14. The test pilots weren't too adept at checking us out, and we found that unless this machine was landed on its main wheels with its tail near the flying position it became very troublesome. At that time, we landed most aircraft, including our Lockheed 10As, in a three-point position; that is, with the main wheels and the tail wheel touching the ground approximately together. At first the new procedure of a "wheel landing" was awkward for us.

I bounced my aircraft badly around the Reno field, sometimes almost digging wingtips into the ground. Slim Lewis was having the same trouble. Furthermore the test pilots would not check us out in stalling the 14—they said that they had tried it once or twice and that was enough. They reported that a stall with the 14 was so violent that the aircraft flipped right over on its back. It was apparent that we were going to have to find out how to land the thing properly and to stall it at altitude by ourselves. We were still unable to make safe landings when our test pilots returned to

Burbank. We were now on our own.

There was little use our staying in Reno; we might as well get under way and practice our landings en route. We decided to fly north to The Dalles on the Columbia River, west to Portland, and north again to Seattle expecting to arrive there just before dark to overnight.

Slim Lewis, with Maurice McGregor, led the way in CF-TCE. Slim made a good take-off, climbed out and headed north. I followed in CF-TCD, with Lindy Rood. By the time we were airborne, Slim had disappeared, but it didn't matter, as we were to follow the same route. Lindy and I were so fully occupied getting used to the new machine, its fuel controls, radio and so on that, once airborne, we were quite busy. Things went well, but in the back of my mind was the realization that eventually I was somehow going to have to land!

Winds delayed us on our leg to The Dalles, and as we flew west along the Columbia, the ceiling was lowering and a light rain was falling. We droned fairly low over Portland, then swung north. Darkness was falling, and by the time we reached Seattle the field was barely distinguishable, though we did recognize Slim's aircraft parked on the tarmac below.

Anxious, I approached the runway, fully expecting the usual bouncing arrival, particularly with the restricted visibility. As the runway flashed below, my main wheels touched and, unconsciously I must have pushed the control wheel slightly forward to hold the wheels on the runway. We didn't bounce but rolled smoothly to a stop and shut down alongside CF-TCE. Slim, who was waiting for us, asked, "What the devil did you do to make such a good landing?" I had no idea, which left us more baffled than ever. We spent the evening in our hotel room with a bottle of whiskey, the four of us trying to figure out why my landing had been good! Slim's had been rough.

The next day, May 12, we took off for Vancouver. On arrival there was a strong crosswind blowing. The Lockheed test pilots had told us always to use 75 per cent of flap for landing, and that if we bounced badly, we must never under any circumstances retract our flaps or we would be in real trouble. We found out the hard way that they were wrong. Our arrival had been well publicized so there was a crowd at the airport, including news and camera men.

Slim landed first, while we circled overhead. We watched him bounce badly as he touched down but he got it under control. I approached next, still trying to figure out what I had done the day before. I touched down, with 75 per cent of flap, fighting the cross wind. The aircraft bounced hard and hopped down the runway like a kangaroo almost out of control. Lindy admitted later that he had visions of the radio direction finding loop, mounted in the nose, coming back right into his stomach. There was nothing left but to apply full power and try going around, in spite of the 75 per cent flap. The aircraft responded well, and as I climbed I told Lindy to bleed the flaps up slowly. There was no difficulty and we circled for another landing. We bounced again but no so badly, then taxied in and joined Slim and Maurice. We had damaged one fuel tank in the first landing but had learned that the aircraft could be taken around again and the flaps slowly retracted. We had not, however, learned the secret of landing the Lockheed 14.

My damaged fuel tank had been sealed by the following morning, so we decided to push on to Winnipeg non-stop to check the fuel ranges of the aircraft. We took off early on May 13 and headed for Winnipeg. The dust storms were so bad across the prairies that at times there was almost a level floor below us composed mainly of dust. We reached Winnipeg after 5 hours and 15 minutes. Our wives and all the company people who could assemble were there to welcome us. Slim went in first and made a reasonable landing. I followed, and again bounced, though not nearly so badly as before. We were starting to acquire a feeling for the aircraft. In another day or two we learned how to land it properly, and from there on there was no further landing trouble. The problem of stalling, however, was still to be overcome.

Ronald George and Howard Sandgathe had been appointed instructors before our departure to Burbank. On our return we found they had done a fine job and our training school work was up to date. The Lockheed 10As were now operating a limited schedule as far west as Vancouver. The fuel range westbound from Lethbridge to Vancouver was sometimes critical due to headwinds and from time to time required a refuelling stop at Oliver, B.C. Only mail and company personnel were being carried at this time.

In Winnipeg our ground crews checked over the Lockheed 14s and had them ready to fly on May 16. Slim Lewis took one and I the other, and we started converting some of the Lockheed 10A captains to the new aircraft. At the top of the list were the two new instructors, Ron George and Howard Sandgathe. Slim and I both knew that we'd soon have to confront the 14's vicious stalling characteristics. Our training program for each pilot included a course of unusual manoeuvres, one of which was a series of stalls, both in the clear and blind. Before that manoeuvre could be carried out with our students, both of us must be fully qualified. Each morning, as we detailed the particular exercises for the day, we knew we'd have to face the evil moment.

One morning late in May Slim looked at me and said, "Lewie, I guess this is the day." I knew exactly what he meant. I simply answered, "Yes, Slim, I guess so."

After setting our students to work at other things, Slim and I climbed into one of the 14s. He took the left-hand seat and I took the right. We started up, warmed up, and took off without saying a thing. At about 12,000 - 13,000 feet Slim tapped me on the shoulder and stuck his right hand out. We shook hands quietly, then Slim started to pull the nose up and throttled gently back. The aircraft felt very smooth, no shudders or warnings, then suddenly flipped halfway over violently. Loose things flew around, but Slim shoved the nose down and she straightened out. It was a pretty rough stall but not as bad as we had been led to believe. He did a couple more, then we changed seats and I did about three. We were as pleased as two small boys. We could now get on with the job of confidently teaching the others. We agreed, however, that the stalls, particularly the left-hand ones, were bad enough that Lockheed should do something about them. As a result, Lockheed, in conjunction with our senior engineers, Oliver West and Jimmie Bain, started a program of tests and alterations a bit later on, which we flew from Winnipeg, and which led to better performance.

During this time, Sandy Sandgathe, while instructing in stalls in the 14s, got into a vicious high-speed stall, fortunately at a safe altitude. It happened that Slim and I were on the ground, watching Sandy at work. We saw the aircraft stall violently and head for the ground, coming straight down and quickly building up speed. We thought they were going to crash, but Sandy managed to pull out at a low altitude and land. When he taxied in, we saw that both wingtips were bent up at right angles. Sandy reported getting into the high-speed or crossed-control stall with the airspeed indicator still showing a fair rate of speed. He had taken normal recovery action but this resulted in airspeed well beyond the allowable maximum, with the nose straight down. Finally he yanked the power right off and was able to pull out; but the force of pull-out at very high speed had folded up both wingtips. Both wings were found much over-stressed internally and had to be replaced, but from this incident we learned how to cope with the high-speed stall.

Now that Ron George and Sandy Sandgathe were instructing, it was possible for Slim Lewis and me to spend much of our time acting as check pilots on the scheduled runs in order to monitor the progress of the pilots. We also managed to do some instructing and test flying as well.

On July 4 we started a system of numbering our aircraft instead of using the registration letters. This simplified things, especially over the radio. The numbers started with 21. Our first aircraft, CF-AZY, became number 21. The numbers progressed through the Lockheed 10As, and then through the 14s.

On August 2 I was advised to do a test flight with W/C A.T. Cowley of the Department of Transport to determine the practical ceiling of the Lockheed 14 operating on one engine with a full load. This was to enable us to get certification for operations over the Rocky Mountains. For the tests, we loaded the aircraft with sand bags. With me in the captain's seat, we shut off and feathered one engine at various altitudes until we were sure of the maximum height that the aircraft would hold safely. Tom Cowley was satisfied and cleared the way for passenger operations soon to begin across the mountains.

We were taking delivery of more Lockheed 14s and gradually replacing the 10As on the scheduled runs. We now started to carry proper air mail rather than ordinary mail taken from trains; in addition we sometimes carried Dominion Government personnel mostly from the Department of Transport.

18

Check Flights and Horror Stories

The airline decided to familiarize its captains with the emergency landing strips along the routes from Winnipeg to Vancouver, up to Edmonton, and later from Winnipeg to Montreal. I was detailed to be the captain of the familiarization flight. On each leg, east and west, I was to carry eight pilots, so that 32 pilots in all would take part. We could not spare a Lockheed 14 from our services so we took a 10A, number 24, registration CF-TCB. I left on the first trip with eight senior captains involved with the western route, including Herbert Seagrim and Lindy Rood, who later became top executives of Air Canada. We did approaches, landings and take-offs at every strip en route.

We left Winnipeg and headed west on August 16, 1938, arriving in Vancouver on the evening of August 18. Along the way we landed at Rivers, Broadview, Regina, Swift Current, Medicine Hat, Lethbridge, Coleman, Cranbrook, Midway, Oliver and Princeton. Taking aboard another eight captains who were waiting in Vancouver, I took off on the morning of August 19, landing at the same fields on the way back. On this journey we overnighted at Cranbrook. After dinner, we were invited on a tour of the local brewery, which concluded with a free beer or two. While we were sipping away, we heard our evening flight eastbound from Vancouver go overhead, and someone remarked that life was more pleasant in the brewery than in the night skies over the mountains.

We arrived back in Winnipeg on the evening of August 20. On the 23rd, with another eight captains, I left Winnipeg again, arriving at Montreal the following night. En route we landed at Caddy Lake, Vermilion Bay, Amesdale, Sioux Lookout, Allanwater, Wagaming, Grant, Kapukasing, Porquis Junction, Earlton, Gillies, Killaloe and Ottawa, all in Ontario.

Another eight captains awaited in Montreal and we left on the 26th, getting back into Winnipeg next evening. When the job was finished, most of our captains had seen the emergency fields they would fly over on their scheduled routes. Fortunately they were rarely to need them.

My next assignment was to act as captain of a radio range test trip from Winnipeg to Montreal and return. This was to ensure that all radio ranges were functioning properly, that their courses were correct, and that their approach legs were in the right place in respect to the airfields. If so, everything would be ready for our eastern division to start operations.

I took the Lockheed 14, CF-TCE, or as it was now known, number 27. On the flight were D.B. Colyer, the chief technical advisor; H.O. West, chief of engineering; O.T. Larson, chief of meteorology and despatch; and Slim Lewis, chief of flying. Although I acted as captain, Slim came up from time to time and spelled me off.

We left Winnipeg on August 30 and were back on September 1, having checked fields at Ottawa, Toronto and Montreal. On the return trip we flew non-stop from Toronto to Winnipeg following the new airways route via Kapuskasing. We did this to check fuel ranges, but were nearly out of gas when we landed. The engine fuel mixture settings had been leaned out as far as possible, and we very nearly stretched this too far. We flew the radio ranges to check for operational suitability.

On return to Winnipeg, there was a series of tests to be done on the Lockheed 14. The first covered all phases of single-engined performances at various speeds, and from an empty aircraft to one fully loaded. Following that, I flew the aircraft in a series of wing de-icer tests, carrying out stalls with the de-icers in operation. I was then sent out to the Lethbridge-Edmonton section to do a thorough night flying check ride with each pilot on that leg, returning to Winnipeg on October 2.

October 4 and 5, 1938, we carried out tests in connection with the vicious stall characteristics of the 14. The engineers had decided that the more vicious left-hand stall was partly caused by two lights on the port wing tip. There was only one on the starboard tip. One of the port lights was the red navigation light and the other was a red warning light. The starboard tip had only a green navigation light. The lights were on the end of a metal extension projecting a few inches from the tip. Apparently serious air flow burbles were taking place there.

Using Lockheed 14, number 29 I did stalls with the lights in normal positions. Lockheed had fastened small strands of red wool all over the upper surface of the wings and mounted movie cameras on the aircraft; and there was a trailing speed indicator below, which we let out on a cable. The action of the wool strips was photographed during our stalls with the lights in position. The lights were then all removed and I did another series of tests. The results showed that the lights should have been flush with the wing surface. The red warning light was removed from the wing altogether and placed in the nose with the radio loop gear. This made the aircraft stall about the same in both directions. From the tests also came the fitting of slots behind the leading edges of the wings, which made the stall much gentler. These wing slots were a boon later in the Second World War, when the Lockheed 14 became the Hudson patrol bomber. Without the wing slots, the toll of pilots and aircraft in training and on operations would have been much heavier.

Throughout the rest of October, I alternated between more test flights, some instructional flying, and a number of flights to and from Vancouver, riding as check pilot with various crews. One of these check flights came very close to being my last.

The night of October 26 I decided to check the flight leaving Vancouver for Winnipeg. The captain was Maurice McGregor and the co-pilot was W.D. "Don" Brady, in aircraft 31. Aboard, in addition to a load of mail, was a small group of DOT officials returning to Ottawa. The weather was expected to dictate an instrument flight across the mountains, with some icing. A *mild* occlusion was forecast near Grand Forks at the time of our passage. Weather forecasters were still learning to forecast mountain weather and, of course, our pilots were also learning to fly it. The *mild* occlusion forecast turned out to be the worst weather we had encountered so far in our night flights across the mountains.

Maurice McGregor took his place in the captain's seat with me, as check pilot, on his right. Don Brady moved back to the cabin with the DOT passengers. We took off from Vancouver at 7:15 p.m. and climbed through cloud and rain to our cruising altitude of 13,000 feet. At this time we didn't have oxygen on our aircraft; that was to come later. As a result, we had to work at staying aware of exactly what was happening when flying at high altitudes.

We crossed over the cone of silence marking the Princeton radio range and were flying blind, encountering fairly heavy icing. From Princeton we swung southeast towards the radio range station in Grand Forks. Static and turbulence were getting worse and the icing heavier, forcing us to climb. We tried 15,000 feet but we were still in heavy ice and static. The wing and carburettor de-icers were working fairly well, but we were having trouble with our propellers in spite of our anti-icer fluid. We frequently had to push our hydromatic propeller levers forward to shake off some of the ice and large pieces would thud against the sides of the aircraft.

Just as we thought we were near the cone of the Grand Forks radio range station, which we had much difficulty hearing through the heavy static, we hit a real belt of turbulence, one jolt flinging the aircraft about violently and causing us almost to roll over. At this moment Maurice was trying to re-tune our main radio receiver, the handle of which was on the roof of the cockpit. In the upheaval the gear train to the tuner of the receiver broke. The main receiver was not tuned now to the station, nor could we alter it further. By coincidence a tube burned out in our standby receiver, causing it to fail. These things we did not know at the time, but were told about later by our technicians. Finally, the only receiving set left for radio range operation was a small emergency one powered by dry cell batteries. We tried for some time to pick up the ranges on our main sets but couldn't. Later we turned on the emergency set.

We knew now that we were in real trouble. The icing was getting worse, so we climbed to 17,000 where it seemed a bit lighter. We also decided to return to Vancouver, as we still feared that the weather was worse ahead. We tried to relocate Princeton but couldn't with the heavy static. Realizing that we were lost, we used our emergency procedure: fly due south seeking an area of more radio ranges in the United States. We checked each other continually and noted that we were both getting a bit blue from lack of oxygen. At times icing forced us up to 19,000 feet. We flew south for an agony of time but could get no response from our little emergency set. Our fuel was going down fast with the extra power used to maintain height. Don Brady came up front to report that the passengers were all out cold and blue in colour. One or two were bleeding a bit. There was nothing we could do about it.

Finally, after an elapsed time of about three and a half hours, battling ice, static, turbulence, and unconsciousness, we decided to continue our southerly course until down to 20 minutes of fuel. Then we would start a descent. We knew that this would mean almost certain death in the mountainous terrain, or that if we were further west than we thought, we might break through off the US coast. Meantime we had been giving an emergency call on our transmitter every little while but had no knowledge that it was being received. We didn't know that the company and DOT had heard our calls and knew we were in serious trouble. They had cleared the air of all other radio traffic within our range and broadcasted a message to people to listen on both sides of the border for the sound of our engines. They had no equipment for getting a fix on us. Maurice and I were having trouble keeping our minds working. We wrote down the course we wished to follow on a piece of paper and held it beneath the figures on our directional gyro. Periodically we would both check it to make sure we were on course and that our minds were still working.

We were unaware that when we turned back at Grand Forks we were in the middle of a powerful mountain weather front extending south of the border over Grand Forks and up to Prince Rupert to the northwest. Winds in the front, from the southeast, reached 150 to 200 mph at our altitude. These figures were provided later by US mountaintop stations. When we turned south, we had not escaped the front, so were still in the high-wind area. Consequently, we were actually drifting backwards to the northwest as our ground speed was less than the winds we were in!

Our fuel supply was worsening when we began to pick up faint radio range signals on the emergency set. When we could finally hear the identification signals of the station, they were "VR" (the identification letters for Vancouver radio range). We didn't believe it, for we were sure we were well down into the United States. The signals got stronger and finally we found that we were on an "on course," that is, on a leg of the radio range. We didn't dare try to orient ourselves due to our fuel shortage, so just held our southerly course. The signals got stronger and stronger, and then we broke out into a big hole in the cloud; through parts of our iced up windows we saw the lights of a big city below. We banked steeply and looked down at the city, but in our condition could not recognize it, probably because we thought we were at least over California.

We both agreed, "We'll go down here and land somewhere." By now the range signals were very loud, but we could not believe we were over Vancouver. We went in and landed, unable to believe our good fortune.

We soon learned what had happened. We had broken out of the frontal winds north of Vancouver, by the grace of God, and, while still flying south, had arrived over Vancouver *from the north.* This was confirmed later by the telephone calls of people living in the mountains who had heard engines where aircraft were not normally heard at night. We were well down on our last tank of fuel and had spent 4 hours and 5 minutes in the air since taking off. We were very lucky people.

After a rest, warm food and a change of radio sets, we were advised that the front was weakening, so we took off again with the same load. After a bit of a battle over the mountains we got through to Lethbridge. Once over the prairies we had no further trouble.

On our return to Winnipeg we learned that the McKee Trans-Canada Trophy for 1938 had been awarded to the operating personnel of Trans-Canada Air Lines for our work in building and flying the early operations of the line. We were naturally very pleased, the McKee Trophy being the top award in Canada for civil air operations. In later years some of us were to be given the award again on an individual basis — Frank Young, Barney Rawson and myself, so far.

I continued to ride back and forth between Winnipeg and Vancouver, and occasionally to Edmonton, as a check pilot, and was pleased to note the skill with which our crews were handling their flights, particularly on instruments or in icing.

On the night of November 17 I went out to Winnipeg airport to watch the daily departure of our scheduled flight to Vancouver. Departure time was 11:30 p.m. The crew was Captain Dave Imrie and co-pilot Jack Herald. There was a load of mail aboard, but no government passengers. Imrie and Herald had gone through our training courses with flying colours and were rated as a first-class crew. I said, "Good night, good trip,'' to them, and they climbed aboard and took off. I returned home and went to bed.

About 4 a.m. the phone rang. I crawled out of bed and answered it. It was Slim Lewis with the bad news that Dave Imrie and Jack Herald had crashed at Regina. The aircraft had burned and both pilots were dead. Lockheed 14, number 28 was made ready immediately for Slim and I to fly to Regina along with some of our engineering personnel. Quickly dressing and swallowing some hot coffee which Lin had prepared, I tore out to the field, met Slim and party, and took off.

At the time of Dave Imrie's crash, the weather was good, though fog patches were forming off the western side of the field where the crash took place. By the time we had flown to Regina, the whole area was locked in a dense ground fog. We flew around but could not get down. Around Moose Jaw, 40 miles to the west, holes were appearing in the fog, so I flew over there and landed. We hired a car, and drove to the crash site.

The burned-out remains of the aircraft were about a mile or two west of the field. Marks showed where the Lockheed had skipped across the ground once or twice before stalling in and hitting the ground hard. The RCMP had roped off the area and had it well guarded.

We went over the wreckage in detail during the next two days. The Department of Transport view was that immediately after take-off Dave had seen the fog, tried to fly under it, and hit the ground. I did not agree — I had given Dave most of his airline instrument training and was sure he would have wanted to climb above the fog. There had to be another explanation for the crash but it was to take another six months to find it.

I had noticed a hand fire extinguisher lying near Dave and Jack and that its handle was extended and bent. That indicated to me that they were possibly using it at the time of the crash. Furthermore it was the extinguisher which was normally located in a metal bracket on the floor behind the last passenger seat, close to the cabin door. It was mainly used by the mechanics and held in readiness while starting engines. There was a second extinguisher aboard in a metal sheath or bucket alongside the pilot's seat, but it was still in position in the wreckage. The question was, if a fire extinguisher was required in flight, why go for the one at the back of the cabin when there was one right in the cockpit? A second mystery arose when we found the wooden lid from a small control box located on the floor between the pilot seats. This control box contained the engine fire extinguisher controls and also the main cross-feed fuel control valve. We found the wooden lid just ahead of the wreckage of the nose of the aircraft and noticed that it appeared to have been burned on its underside only. For these reasons I believed there had been a fire of some kind in the aircraft and that the crew had tried to put it out, but I couldn't prove it.

A little later on, the engineers advised us that the cockpit fire extinguishers were all stuck in their metal buckets and couldn't be pulled out. Slim Lewis and I checked all the aircraft in the shops and verified this. When the extinguishers had been fitted in the buckets, a green baize lining had been glued in to stop the extinguishers from rattling. The extinguishers which had been put in the buckets before the glue was dry, stuck fast! We knew now why the cockpit fire extinguisher in Dave Imrie's crash hadn't been used.

A month or two later, North West Airlines had a fatal crash of one of its Lockheed 14s during a night take-off at Billings, Montana. The conditions were almost identical to ours. We checked with North West and found that the lid from the control box containing the fuel cross-over valve was also found just ahead of the aircraft, and it also was burned on its underside.

This was sufficient evidence to have Lockheed look into the matter of the control box. It was found that the electrical switch in the control box, which turned on a light whenever the lid was opened, was not vapour-proof. The fuel cross-feed valve tended to leak fuel slightly into the box and under the right conditions a fire could ensue. We knew now that Imrie and Herald had been fighting a cockpit fire during the last stages of their take-off. Lockheed removed the live fuel lines running under the control box and had the cross-feed valve operated remotely. There was no further trouble of that nature from then on.

At this time, we decided to look into a new synthetic training aid, the Link Trainer, which had just come to Canada, owned and operated by the RCAF at Trenton. I was to take a course on the Link to determine whether we should buy one or two. Lin and I departed for Trenton, seen off at the station by a group of senior pilots, who had sprinkled rice throughout our baggage and showered us with the stuff. On the train we explained to the porter that we had been married for years but didn't succeed in convincing him. The course at Trenton was conducted by my friend F/L Frank R. Miller. He gave a good course and as a result we ordered some Link Trainers and used them for instrument and radio range training. Later on, this section was placed under the control of Pat Howard.

The author (in classic attire) at Lethbridge in 1928 while flying a de Havilland Moth belonging to Southern Alberta Airlines. (Below) The author owned and operated CF-APQ, the only Command-Aire 3C-3 ever registered in Canada. Ernie Boffa ferried it to Medicine Hat in 1931. Later that year Leigh sold it and moved to the East Coast. APQ was known to be operating in Alberta as late as 1936.

Two fine views of CF-AHE, one of the Fokker Standard Universals operated by Maritime and Newfoundland Airways in 1930. In the bottom view the author and his engineer are on their visit to Newfoundland and are attracting considerable attention from the local outport folk. AHE had a silver fuselage and varnished wooded wings.

(Above) Lin Leigh at Sydney with AHE. (Left) The author in 1932 with his fellow pilot on the North Sydney operation, George Silke, who was a World War I aviator. (Below) The author, left, and Bobby McGowan, right, after a test flight at North Sydney just before AHE and AHJ were ferried west.

The Blind Flying Course at Camp Borden in 1932 included (back row, from the left) Charlie Gordon, Bill Page, Ralph Milani, Des Murphy, Lewie Leigh, George Silke, Bill Dean and Dinny Dinsmore. In front are Gordon Steves, Jeff Home-Hay and Bill Catton.

AHJ at Cooking Lake outside Edmonton. It had just arrived from Fort McMurray on a mercy flight.

Fort McMurray around 1930. The buildings on the left were later burned down.

When the Leighs first moved to Fort McMurray, Lin, shown at the front door, made this former butcher shop into their home.

Canadian Airways' rugged Junkers W-34 at Fort McMurray. At the time, CF-ARI was Walter Gilbert's machine, but he occasionally had his friend, Lewie Leigh, do the flying. (Below left) Canadian Airways' chief pilot down the Mackenzie in 1934, Walter Gilbert. (Below right) Trader Dick Halcrow, who had a post on the Barren Lands, meets with some of the tundra natives.

Stan McMillan was captain on Mackenzie Air Service's Bellanca Air Cruiser which freighted pitchblende ore from Great Bear Lake into Fort McMurray in the mid-1930s.

Engineer Bill Sutherland (left) and his pilot, Lewie Leigh, March 1933.

Rudy Huess's Fairchild through the ice at Fort McMurray—not an uncommon sight in bush flying days.

Canadian Airways' Fairchild 71C on the ice at Cameron Bay, N.W.T., in 1934. The drums of aviation gas were brought in by barge the previous summer.

The author in two photos taken at the Boeing School at Oakland, California in 1936. The aircraft is a Boeing 40B-20. Its blind-flying hood arrangement shows clearly.

(Below) Canadian Airways' Laird LC-2B-200 blind-flying-equipped aircraft seen at Columbia Gardens, near Trail, B.C., in 1936.

The author's first class of instrument flying students at Canadian Airways, Winnipeg, in 1936: Paul Davoud, David Glen, Stan Wagner, the author, Jock Barclay, Ron George.

Factory fresh, Canadian Airways' sleek Lockheed 10 photographed in 1936, with company mechanic Al Dyne by the wing.

(Left) Lewie Leigh and fellow TCA pilot Slim Lewis by a Lockheed 14 which they had just delivered to Winnipeg, summer 1938. (Right) The author and his first officer, Ted Allan, at Lethbridge while on an early scheduled TCA flight.

(Above) A TCA Lockheed 14. From this tough little airliner evolved the wartime Hudson patrol bomber, which the author also flew. (Left) The wreck of TCA's Lockheed 14 in which pilots Dave Imrie and John Herald died.

A typical scene at TCA's Winnipeg base. At this time the company's L.14s still had the Pratt and Whitney Hornet engines. These were later replaced by Twin Wasp engines.

(Above) The author and three colleagues beside a Lockheed 10. From the left, mechanic Charlie Grogan, pilot Jock Barclay, the author, and aviation pioneer Bill Straith, who had built his own aircraft in 1912. (Right) CF-TCI after its hair-raising emergency landing at Vancouver.

Members of No. 13 O.T. Squadron at Patricia Bay, February 1941: F/O D.Z.T. Wood, F/L J.L. Gray, F/O W.B. Purvis, F/L J.K. Lynch, F/L A. Vanhee, S/L R. Briese, F/L Z.L. Leigh, F/L C.C. Austin, F/L H.J. Winny.

(Left) The author's longtime friend, Air Commodore John Plant, at the time commander of No. 9 (T) Group at Rockcliffe.

(Below) Fortress 9204 about to depart on No. 168 Squadron's first overseas mail flight.

(Above) No. 168 Squadron's first overseas flight about to leave Rockcliffe, December 15, 1943. From the left are G/C J. Sharpe (back to camera), A/V/M Nairn, A/M Leckie, W/C Findlay, S/L Middleton, Hon. C.G. Power, W/C Leigh.

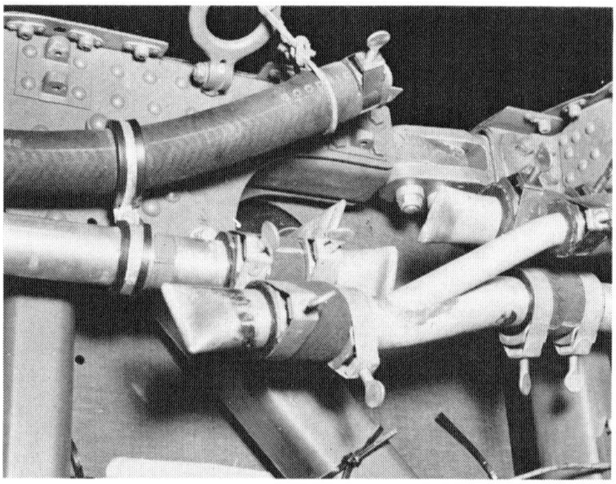

The crimped fuel lines that nearly spelled disaster for 9204 on its first operational flight; and the rubber inspection cap showing the hairline crack that saved the day on the author's return from that trip.

The author, Lin Leigh, Field Marshal Montgomery, Fran Cathcart and Col. H.M. Cathcart during Monty's Canadian tour in September 1946.

The Leighs, at Goose Bay during Lewie's posting there as CO.

Some winners of the Trans-Canada Trophy and James Mollison and A/C John Fauquier who attended the official presentation of the trophy to Barney Rawson at the Fort Garry Hotel, Winnipeg, on October 2, 1948. Barney Rawson (extreme left) is being congratulated by James Mollison. First row behind Mollison: Punch Dickins, J.A. Wilson, Romeo Vachon, Pat Reid, Moss Burbidge and Murton Seymour. Second row: A/C John Fauquier, Lewis Leigh, Dan McLean, Doc Oaks, George Phillips, Tommy Siers and Wop May.

The author at home in Grimsby, Ontario

19

End of the Line with TCA

The next month or so was one of alternating between periods of instructing and riding with various crews on the scheduled runs as check pilot.

On January 4, 1939, a special test flight almost caused the grand finale for Lockheed 14 number 29, plus Jimmie Bain, the senior engineer, and myself. We had found in our tests, and pilots who had encountered heavy icing on their routes had reported, that in spite of full use of anti-icing fluid on the propellers, the blades iced up heavily and caused serious vibration and loss of efficiency. The Engineering Department had a theory that the engine cowlings were causing a peculiar air flow pattern ahead of the engines and that this caused the fluid to be blown off into space by the time it had flowed half-way down the blades. They removed the cowlings on both engines of number 29, filled the anti-icer tank with a coloured dye, and then asked for an air test to show the flow of the dye in the air at cruising speed without engine cowlings.

I wondered what the uncovered engines were going to do to the air flow but was assured that it wouldn't be too serious. In any case, Jimmie Bain, a very good friend, was going to ride with me as co-pilot to monitor the operation closely. I opened the throttles and we started rolling down the runway but didn't seem to be picking up speed as quickly as we usually did. By the time we had reached take-off speed, we were almost at the end of the runway. I pulled back hard on the controls and managed to get off the ground, but I knew then by the feel of the controls that we were in trouble.

Our speed did not increase appreciably and a terrible vibration set in on the controls until finally the whole aircraft was quivering. Jimmie Bain went back into the cabin to have a look and returned in a hurry, saying, "Get the damned thing back on the ground — the tail is jumping up and down badly.'' I couldn't gain much altitude or speed and found that when I tried to turn, I seemed to be on the verge of a crossed-control type of stall. We staggered around the circuit slowly and shakily, both of us almost holding our breath that everything would stay together. Finally I snapped down on the runway. In our concern we hadn't even remembered to turn on the coloured anti-icer fluid, so no pattern showed on the blades. Dye or no dye, we decided firmly that there would be no further tests with the engine cowlings off. It was clear that the uncovered engines had caused serious burbles in the air flow which had a pronounced effect on our tail control surfaces. In the end, a grooved rubber strip was developed which, cemented to the leading edges of the propeller blades, helped with the distribution of the anti-icing fluid.

We were now becoming a real airline and were making plans to get our first paying passenger flights on a transcontinental basis under way by April 1. Stewardesses were being hired and trained under the control of the senior stewardess, Lucille Garner.* On our early flights, the girls had to work in the cabins without oxygen, which was particularly difficult on the mountains. They were a good group, all registered nurses, who did their best, sometimes under trying circumstances.

Slim Lewis and I did more and more checking of pilots on the scheduled trips in order to have everyone in tiptop shape for the beginning of our passenger flights. In March Lockheed sent their Lockheed 14 experimental aircraft No. 1 to Winnipeg to show us the new wing slots, which were to reduce the viciousness of the stall. Having done so much of the original stall testing without slots, I tried out the experimental aircraft. It was very satisfactory, with the stall reduced to a manageable, reasonable manoeuvre. From then on all our aircraft were converted in turn. The pilots were happy except that sometimes, under heavy icing, a projection of ice would form under the slots, posing a minor problem.

On April 1, 1939, our flights went into operation with paying passengers in addition to air mail. Flights left Vancouver and Montreal for Winnipeg on the 1st, then from Winnipeg, flights left for Vancouver and Montreal on the 2nd. I was captain on the first flight from Winnipeg to Vancouver, with Gil McLaren as co-pilot and Margaret Wilson as stewardess. The weather over most of our routes was unusually poor during our first flights, so that we all had trouble or delays.

On my flight west we landed at Regina and Lethbridge without too

* The first group of stewardesses included Kilby Harding, Pat McNamara, Florence Shanahan, Margaret Wilson, Annette Brunelle, Norah Wallace, Sheila Neill, Pat Eccleston (the second stewardess hired), Lela Finlay, Geralde Brunelle, Margaret Beeber, Rose Crispin, Constance Haibeck, Dorothy Price, Evelyn Allan, Marcelle Levac, Margaret Brass and Ruth Leslie.

much trouble weatherwise, but in the mountains the "met" was very bad. I left Lethbridge on schedule westbound but had to return after encountering icing as high as 17,000 feet. I tried again a few hours later and got halfway across, but icing beat me again so we returned to Lethbridge. On the third try we managed to get through, as conditions were improving. The flights all straightened out within a day or two, and from then on we became accustomed to looking after airline passengers.

Among my passengers on the first flight was Robert E. Day, president of the Bulova Watch Company of Canada, from Toronto. During the flight, I invited him into the cockpit to see the view from there, which he apparently enjoyed. A few weeks later, a parcel arrived for me containing a gold wrist watch, a Bulova of course. It was engraved to commemorate the first passenger flights, and the cover of the box told the story on its silk lining. I still wear the same watch today, over 30 years later. It has covered most of the world and been through the Second World War and the Korean War.

Thereafter, I stayed on the run as a captain or check pilot until the end of June 1939, when I became a captain on scheduled operations. My instructing and checking of pilots was finished. I was replaced as chief pilot by instructor Ron George. He now became chief of flying, under Slim Lewis, with me as senior TCA captain. This was a good arrangement on the part of the company because I was not a very good executive type and loved flying better than anything else. Ron George became the executive. I was happy flying my regular night run to the West Coast and back, for it gave me much more time at home with Lin. We purchased a new house further down Borebank Street and bought a new Dodge car. Our pilot group formed the first Canadian Airline Pilots Association in Canada, with Jock Barclay as first president and myself as vice-president. The same organization exists today on a much larger scale.

Our aircraft were fitted with oxygen equipment during this period. We inhaled through a small nose mask which left the mouth clear to use the microphone. At first the pilots were against using oxygen as we felt that we were doing all right on our flights over the mountains, even though we became fairly blue (particularly our finger nails) on each high trip. Gradually, however, we realized that we felt much better and our minds were clearer when we used oxygen. From then on, when we reached the 10,000-foot level, we would religiously strap on our oxygen masks. The passengers also appreciated it, I am sure.

On the night of August 2, I was over the prairies at about 9000 feet between Lethbridge and Regina in aircraft number 30. The co-pilot and I were sitting comfortably with the cockpit instrument lights turned well up. We often did this and flew on instruments even on clear nights, as it was almost easier than flying contact. Suddenly there was a heavy thud through the aircraft and the controls almost jumped out of my hands. Then everything steadied again and felt normal.

We called Regina and reported the incident and that I would make a very careful approach in case there was damage to the aircraft. Regina replied that there would be engineers on hand to look things over as soon as I landed. After a normal landing the engineers inspected the aircraft with flashlights and found we had hit a large bird in flight. One propeller was bloody and some of the remains were jammed between two of the cylinders in one engine. Another large piece had apparently been blown into the tail, damaging the leading edge of the horizontal stabilizer and causing the controls to jump. Some feathers were later sent for examination and we learned later that as far as could be determined the bird was an eagle. Why it was flying in the dark at 9000 feet was a mystery.

On September 2 I started two weeks of leave. Lin and I had not had a really good holiday for a few years, so we started out in our new Dodge. After some wandering around, we finally toured Yellowstone Park and then spent some time in Sun Valley, Idaho. The war had just started and I tried to get a picture of what was happening from American newspapers and radio. Since leaving Camp Borden in 1932, I had been on strength with the Reserve of Officers, RCAF as a flying officer, and it seemed that if I were not called up, I should at least offer my services to the RCAF. After returning to Winnipeg, I went back on my night run to Vancouver.

A little later on, I had a call from some bush pilot friends in Edmonton. They were considering making a trip to Vancouver during the freeze-up period to visit the senior RCAF officer on the West Coast, G/C G.O. Johnson, and to volunteer for the service. I decided to go with them. We were all to meet in Vancouver, though some of them joined my flight in Lethbridge and we flew to Vancouver together. G/C Johnson noted that we were all offering our services and said we would be advised when we were needed. Afterwards, we got ourselves inebriated in a proper and fitting manner to celebrate the occasion, then returned to our jobs to await results.

Our flights were now becoming more routine. Weather forecasting and radio procedures were better, and our flying through heavy weather, particularly on our mountain run, was steadily improving. This situation ended abruptly for me on the night of January 26, 1940.

I had arrived in Vancouver on the early morning of the 25th through some unpleasant mountain weather. Two flights eastbound had been held up by weather in Vancouver, with the consequent backlog of mail and passengers. On the night of the 26th the weather was even worse and I decided I would not leave Vancouver either. At this point I received a long distance call from Winnipeg asking me to do my utmost to get through that night. Apparently the company air mail pay might be affected on the flights that were held up. If I could get through, everything would be all right. I reported that the weather was really bad but that I would give it a try. (Nowadays that type of operation does not occur.)

I took off that night in aircraft number 31 with six passengers and a stewardess aboard. We climbed out in cloud and icing until we reached cruising altitude, but somewhere near Princeton the icing and static became much worse. Our heated pitot head, which gave the airspeed and rate of climb indication, failed due to excessive ice, even with the heat on. The aircraft was coming close to a stall condition with its load of ice so I decided to return to Vancouver. This was a bit difficult without the help of the airspeed and rate of climb indicators, but we finally managed to get back to the Vancouver area after a pretty rough battle. As we lost altitude our airspeed and rate of climb indication returned, the warmer temperatures melting the ice on the pitot head. We were still flying on instruments with no breaks in the solid overcast.

Vancouver control now informed me that the field was closed in by weather and that Seattle and other nearby fields were also closed. There was no aerodrome at Patricia Bay on Vancouver Island at that time, so we were in a bad spot. We couldn't get over the mountains, and there was nothing open on the Vancouver side. The forecasters thought this condition would last through the night, so I decided to try landing at Vancouver while I still had plenty of fuel to work with. The despatchers on the ground agreed.

After ensuring that my aircraft was free of ice and that everything was ready, I carried out the instrument landing procedure. On the final leg into the field, I stayed at the controls and flew by instruments as the co-pilot watched for signs of the ground, our regular procedure. I can recall letting down lower and lower with the landing lights on, ready for a quick landing. The Kollsman altimeter was showing about 40 feet when suddenly the co-pilot shouted, "The deck!" I pulled back on the elevators but was not fast enough. We hit the ground so hard that we were both stunned momentarily. I opened the throttles wide and tried to regain control. We had just one look at the ground, and then after our very hard bounce we were back in the overcast and blackness again. The top of a brick chimney rushed by in the glow from the landing lights. We were very close to the ground.

Finally our heads cleared and we did some checking. The flaps were full down and we couldn't raise them. The landing gear green lights were out. We knew that our undercarriage was severely damaged and our hydraulic system was smashed somewhere beneath us. I advised the ground control of the situation, telling them that I was going to make another attempt to land as soon as possible, as I was forced to run the engines almost wide open. Their exhausts were glowing like beacons through the opening in the cowlings. Trouble with an engine would have been fatal at that point.

We went through the instrument landing procedure again, dragging our broken undercarriage and with flaps stuck down. Ground control advised that fire engines and ambulances were ready, so we began another descent, this time knowing it was going to be final, one way or another — the machine could not go around again.

The descent continued and we turned on the landing lights again. I instructed co-pilot to be ready to throw the master switch, cutting both engines, as soon as I gave the order. Suddenly we saw the ground together, and we were almost on it. I pulled back and called, "Cut." The co-pilot cut the switches and we were on the ground, sliding across the field in a gentle turn. After stopping, we could still only see a short distance. The stewardess and passengers all left the aircraft, followed by the co-pilot and me. We were thankful that there had been no fire.

The ground vehicles couldn't find us at first, but by shouting a bit they located us and took us to the airport administration building. Our attempt to get the mail through that night was a complete failure. One of the engineers who answered the emergency call that night was my brother Dick. He had lately been transferred from Lethbridge to Vancouver.

We were driven down to the crew rooms in the Vancouver Hotel, where we finally went to bed, completely exhausted. After a short rest, I woke to find my heart pounding madly. It eventually settled down, and later I realized that this was the result of shock. I had also damaged my back a bit when we hit the ground the first time. It has been troublesome ever since, coupled with further aggravation to it during the war.

We went back to the airport next morning, but the weather was still bad and nothing was moving. The ground crews were trying to determine how to move my aircraft into a hangar without further damage. I received instructions to remain in Vancouver until Ron George arrived. The following day we went over the whole thing together, including examining the ground marks where we had hit the first time. I flew back to Winnipeg as a passenger with Ron.

In Winnipeg we had a meeting and Ron informed me that he intended to ground me for two weeks for damaging the aircraft. I said that I considered it very unfair as the only reason the accident had happened was my trying hard to help the company. He was determined to ground me, something I adamantly refused to accept. Slim Lewis, feeling badly about the matter, tried to calm things down. Finally I resigned rather than accept such unwarranted discipline.

My days with Trans-Canada Air Lines were finished, but I have happy memories of our work in building the airline, and of the fine people in it. I still consider it to be the best airline, operationally, in the world, and its senior officials and senior pilots are among my best friends. It was unfortunate in a way that I could not see eye to eye with Ron, but in another way it was good, as I really was interested in getting into the RCAF again now that the war had begun.

20

Ops with EAC

I decided to telephone A.T. Cowley, a Regular RCAF officer on loan to the DOT. Knowing Tom well, I felt he could get the information I wanted. I told him of my volunteering for duty through G/C Johnson on the West Coast, that I had resigned from TCA and was ready to go as soon as the RCAF needed me.

He called me back shortly and said I would receive orders to go to No. 11 (Bomber Reconnaissance) Squadron on the East Coast. They had just been re-equipped with Hudsons, the military version of our Lockheed 14. My experience would help greatly in getting the squadron fully qualified on the Hudson. I was to go to the Recruiting Depot in Winnipeg, have a medical examination and sign the necessary papers. The Depot would be advised that I was coming and that my orders would come through later.

I let Marlowe Kennedy of TCA know of my coming Air Force medical and he volunteered to go too, as he was hoping to get back into the RCAF. This was arranged with the Winnipeg Recruiting Depot. Marlowe and I reported to the Depot where everything went as planned. I was accepted almost immediately, but Marlowe had to wait for "leave of abscence for war duties," since he was still employed by TCA. He joined a few months later. He, Bruce Middleton and I were the only TCA pilots who got into the RCAF after the war started, as the TCA boys were frozen in their jobs for the duration of hostilities (though some were seconded for other jobs, such as ferrying). I was glad to be in the Air Force again. Con Farrell of Canadian Airways, Stan McMillan of Mackenzie Air Services and Chuck Austin of Austin Airways were already in. I was pleased to join them.

We rented our recently purchased home to Art Anders, a TCA pilot, and stored our furniture. A Winnipeg tailor started work on two RCAF uniforms, plus a raincoat and greatcoat.

The TCA pilots threw a going away party for me in the Marlborough Hotel, a fine bash at which we all got very mellow. I was presented with a silver beer stein on which are the signatures of most of the TCA pilots at that time. In the middle of the party, two RCAF Regular Force officers (S/Ls Henry M. Carscallen and R.C. "Buster" Gordon) joined us. They were in the area in connection with the towing across the border of aircraft purchased from the USA, since the Americans were neutral at that time. I had the pleasure of working many times later on with Cars and Buster.

My orders finally arrived. I was to be promoted to the rank of flight lieutenant immediately and travel by train to Halifax. There I would report to the Air Officer Commanding, Eastern Air Command. My date of transfer from the Reserve of Officers to active duty was February 29, 1940. In our Dodge, Lin and I and our springer spaniel, Mike, drove out to Medicine Hat, where Lin was to stay with her parents until we could determine our next move. There I said goodbye again to Lin and boarded an eastbound train.

In Halifax I reported to Eastern Air Command headquarters and was ushered in to see the Air Officer Commanding, EAC, A/C N.R. "Andy" Anderson, a former acquaintance as well as an officer and gentleman of the highest order. His instructions from Ottawa were that I was to be posted to No. 5 (BR) Squadron, on Stranraer flying boats. He couldn't understand the change from No. 11 to No. 5 but to No. 5 Squadron I must go. However, it suited me as I would enjoy getting back to water operations. Andy suggested I get a room in the Lord Nelson Hotel as the officers quarters at Dartmouth Air Station, where my squadron was located, were filled up. An RCAF launch went back and forth across the harbour mornings and evenings carrying many officers and men who lived on the Halifax side. I duly took a room at the Lord Nelson.

Next morning found me on the launch crossing the harbour with a group of other RCAF personnel. Once ashore I tracked down No. 5 Squadron headquarters and reported in. I was taken in to meet my new Commanding Officer, S/L A. Dwight Ross, another old acquaintance, who welcomed me and showed me around the unit. My first duty was to get checked out on the Stranraer.

On March 21 Dwight Ross was the captain of a Stranraer detailed to do an inner anti-submarine patrol with a convoy of ships. I went as co-pilot. Dwight explained the workings of the inner anti-submarine patrol as we flew it and let me get the feel of the aircraft. The next day he took me on an outer anti-submarine patrol and again showed me as much as possible. A few days later he moved me into the left seat and flew with me while I made some landings and take-offs. The feel of the water was coming back and I was really enjoying it.

On the 27th I went out on patrol with S/L R. "Dick" Briese as his second pilot. On the return from the patrol we landed on a lake on Sable Island and watched the famous wild horses running around, then took off again and landed in the harbour at Dartmouth. Briese was transferred to the West Coast a bit later on.

The next day I was detailed to do practice bombing with F/L Len Birchall.

Len handled the Stranraer like a dive bomber and was very accurate with his practice bombs. I tried a few but found I had a lot to learn. Len Birchall later was to be called the "Saviour of Ceylon" for his courage in warning the defenders of Ceylon of the approach of a big Japanese fleet. He was shot down and taken prisoner. F/L Fred Carpenter, who in later years became an air vice marshal, was a flight commander in No. 5 at this time.

Suddenly I was notified of my immediate transfer to No. 11 (BR) Squadron, flying Hudsons from Dartmouth. No. 11 was commanded by S/L A. "Jaggs" Lewis. Apparently Air Force Headquarters had realized I had been posted to the wrong squadron. Actually I was quite happy with Dwight Ross and his Stranraers but it was not to be.

Lewis was a friend from my previous short spell at Camp Borden in 1932, so it was nice to see him again. He was a good pilot and an enthusiastic one. Jaggs liked to let someone else do his paper work while he flew, but we had a first-rate squadron and one which really worked hard.

In the next hangar to ours was No. 10 (BR) Squadron, flying Douglas Digbys. In it were S/Ls Buster Gordon, Henry M. Carscallen, and C.F. "Bud" Newcombe, an ex-TCA pilot of the early days, and F/L Clare L. Annis. At Eastern Air Command headquarters was S/L Walter A. Orr. We were to become good friends and cross paths as the years went by. Carscallen was also living in the Lord Nelson Hotel on the third floor and, since we both found it a bit expensive, we decided to team up. We took one room with twin beds, which saved us some money.

Jaggs Lewis set me to work immediately checking out the younger pilots on the Hudson, as well as carrying out some anti-submarine patrols. I also managed to get some time at practice bombing and using my Browning front machine guns.

The Hudson was basically the same as the TCA Lockheed 14 except that the engines were Wrights instead of Pratt and Whitneys; also there was a bomb bay compartment enclosed by hydraulically controlled doors. Two Brownings were mounted in the nose. The navigator/bomb-aimer worked in the nose.

Most units at this time were short of various supplies. We had no rubber dinghies for our Hudsons. They were supposed to be mounted in a special compartment in the cabin door, but we just stuffed the compartments with brown paper. Eventually dinghies did arrive, but we did a lot of flying over the Atlantic without them. A forced landing at sea, particularly in the winter, meant a quick death unless help was immediately available. A man floating in a life jacket wouldn't last long. Our radio equipment consisted of a small prewar transmitter and receiver worked by a wireless operator using a Morse key. In any case wireless silence was rigidly enforced so that the operator spent his time listening rather than transmitting.

Our bomb load consisted of four semi-armour-piercing 250 pound bombs. This was well before the days we had depth charges. If we had to use the bombs, it meant that we must be very accurate; otherwise they wouldn't do much harm. On one occasion F/L R.C. "Rip" Ripley, returning from convoy duty, found that one of his bombs had dropped from its carrier and was rolling around in the bomb bay. He landed carefully and the armourers had a touchy job easing the bomb safely down through the partly opened bomb-bay doors.

About this time Jaggs Lewis had the idea that Hudsons could be ferried over the Atlantic to England if a special cabin fuel tank were fitted. This would avoid dismantling them for shipment by sea. He and I scouted around and found a long metal fuel tank from a damaged aircraft. After plenty of head-scratching, our engineers managed to get the tank into the cabin of one of the Hudsons. We located an extra outlet on the fuel selector valve which let us connect the tank to the selector with a long piece of rubber fuel hose. We then asked for permission to do a trial flight to the United Kingdom. Senior officers came from Ottawa, stopped us and made us remove the tank. Later on, Hudsons were ferried across the Atlantic, with proper long-range tanks, of course. About this time, F/O W.F.M. Newson (later an air commodore) became my usual second pilot-navigator Bill was the son of Assistant Commissioner Newson of the RCMP and a very enthusiastic airman.

In June, we were pulled off our squadron duties and ordered to fly Brigadier Page of the Canadian Army to Ottawa, leaving on June 5. At this time the United States was still neutral so that legally, when flying a military aircraft, we had to fly all the way around the northern borders of Maine in order to get to Montreal or Ottawa. The civilian airway went straight across Maine, saving a lot of flying time. Having recently left the civilian aviation world, I decided to use the civilian airway across Maine regardless of regulations. We climbed to a good altitude so that our aircraft numbers couldn't be read from the ground and just flew directly across. We returned from Ottawa on June 7 without Brigadier Page but with S/Ls J.G. Bryans and Ted Luke, who were going on business to the East Coast.

As we approached Maine on our return, the clouds became a solid overcast. I climbed high again, on top of the cloud layer, and headed right across Maine again. The senior officers in the rear were suspicious that I was taking a short cut, but with my tongue in my cheek I said I wasn't. On landing at Dartmouth, however, they were sure we had come directly across Maine as our flying time was too short for the trip all around the state to the north. After a bit of discussion they chuckled and let the matter drop.

Temporarily stationed at Dartmouth at this time was No. 1 (Fighter)

Squadron with Hurricanes under the command of S/L E.A. "Ernie" McNab. The unit had been ordered overseas on a troop convoy which was beginning to form up. A few days before they left things got quite hectic on the third floor of the Lord Nelson Hotel. Farewell parties were organized and Henry Carscallen and I decided to have some of our friends in for a drink. The room was so full there was standing room only, but they were all having a good time and the drinks were flowing freely. When they had all gone Cars and I looked at each other and at our room. The boys left their signatures all over the walls, the bedspreads and anywhere else they could find space to write. We had visions of a hefty redecorating bill so we avoided the manager, Mr. Forbes-Thrasher, as much as possible, particularly after he had had our room straightened out. Finally, after days of dodging him, we got up enough nerve to ask him how much the bill was. He looked at us and laughed, knowing full well that we had been avoiding him. "You don't owe anything," he said. "Don't you fellows know there is a war on and these things will happen?" We thanked him and went to our room in great relief for a drink or two.

The troop convoy carrying No. 1 Fighter Squadron and part of No. 112 Squadron, an Army Co-operation Squadron from Winnipeg, was awaiting orders to sail. Our Station Commander, whom I had known in prewar days, was R.S. "Bill" Grandy, a fine pilot and officer who had just been promoted to group captain. He asked if I would like to go on board the troopship that evening to see Ernie McNab and the boys once more before they sailed. We knew they were going to sail next day because No. 11 Squadron was going to be involved in the air protection of the convoy.

We boarded the troopship with Grandy wearing his new gold-braided cap for the first time, had a few drinks with Ernie and his boys, and then heard that two of our old Canadian Airways bush types were down in the bowels of the ship, going overseas with the Winnipeg unit. Bill and I set off to find them, going further and further down in the ship, stepping over troops sleeping in the corridors in places. Finally we found a little cubbyhole a long way below decks and found our old engineer, S.A. "Sammy" Tomlinson, and wireless operator Freddy Searle. We had another drink with them, wished them luck and got back on deck. As everything was sealed up for sailing, the armed sentries on the deck wouldn't let us off the ship. Bill did his best to convince them that we were supposed to get off but they wouldn't listen, until a senior officer came along and cleared the matter up. Back on the dock Bill Grandy took off his gold-braided cap and said, "What the hell good is a thing like this if it won't even get us off a ship?"

June 11, 1940, was a long day of flying. I took Hudson 772 and F/Os Louis Dubuc and Teddy Williams on an early morning patrol to escort two Dutch warships that had escaped the Nazi invasion into Halifax. A little later in the day, with other No. 11 aircraft, I took off in Hudson 771 with Bill Newson to escort the troop convoy carrying our friends. We stayed with them for five hours and then had to return for fuel. By the time we were ready again it was dark but we took off to see if we could do a night convoy job. We found the troopships and stayed with them for an hour, after which it was very difficult to locate them each time we returned to them after patrolling ahead. We finally had to return to base, very tired indeed.

Lin arrived in Halifax, having driven our Dodge all the way from Medicine Hat with her mother, and with our dog Mike on the back seat. She had done a grand job coming that far as she had to do all the driving herself. We stayed temporarily at the hotel, giving up the joint quarters with Henry Carscallen. After a few days her mother returned west by train and Lin and I moved into a couple of furnished rooms near the harbour. It was good to be with her again and to eat some home-cooked food.

No. 11 was flying a lot of anti-submarine patrols with convoys during these days. The convoys were huge and unwieldly, sometimes numbering up to 125 vessels. Later they would be reduced in size, making for better operations. We spent a lot of time, it seemed, flying low over the water in poor visibility but managed to complete most of our operations. No. 11 Squadron was a hard-working, high spirited unit.

On June 21, with F/O Teddy Williams as navigator, I was instructed to patrol out to sea and intercept two British warships, HMS *Furious* and *Devonshire.* In command was Vice Admiral Cunningham, who later became famous in the Mediterranean. We located the ships, which were steaming fast for Halifax. Our procedure at that time was to approach the ships at about 1500 feet and hold a steady course towards them. The ships would then challenge the aircraft by flashing the code letter of the day and immediately the aircraft would flash the correct answer of the day. The aircraft would then begin to escort the ships without danger of being shot down by them. On this occasion we carried out our procedure, approaching and waiting for the signal from warships. It didn't come, so finally I told Teddy to flash the letter of the day. It was answered immediately, but we were almost over the ships. We did not know it, but the procedure in the United Kingdom had been altered, with the result that the ships had everything ready to fire at us when we flashed our letter. This all came out a few days later.

We worked with the *Furious* and *Devonshire* and finally they were safely docked in Halifax. Vice Admiral Cunningham came ashore to attend conferences in Halifax, at one of which our challenging procedures were altered to conform with the overseas pattern.

I was ordered to take Hudson 772, with second pilot-navigator Bill Newson, fly Cunningham and his aide, Lieutenant-Commander Paul, to

Ottawa for a conference, and then bring them back the same day. Once we were on course I asked Bill to invite the Admiral up to the cockpit. He sat in the folding seat at my side. We talked for a while, then suddenly he said, "Who was the damned fool flying the aircraft which intercepted us a few days ago? He didn't challenge us until he was nearly on us. I was going to give the order to shoot when his signal showed." I said, "I am afraid, Sir, that I was the pilot, but our procedures were the reverse of yours. Now ours are the same as yours." The Admiral then said, "It was pretty close. We had your speed and altitude registered accurately so you wouldn't have had much chance." I asked, "What altitude did you register, Sir?" He said, "About 4000 feet." I did not tell him that we were flying at exactly 1500 feet at the time. We had a pleasant trip and flew him back to Halifax safely that same day.

Normal operations went on until July 5 when I was sent out to our gunnery range off the coast to get in some practice with my front guns. F/O "Mike" Michalski was my second pilot-navigator. We flew into position for firing and, at the proper moment in the dive, I released the safety catch and pushed the firing button. There was a blinding flash with fire and smoke through the aircraft. I had to pull up to avoid the water as the cockpit was filled with smoke from a fire which had started in the ammunition bags in the nose of the aircraft. I opened my side window a bit to clear out some of the smoke. Mike grabbed a fire extinguisher and went into the nose, where he managed to put the fire out, a courageous thing to do as bits of metal were still flying around. When eveything was under control we flew back to base. Examination revealed that the port gun had burst and set fire to some damaged ammunition. The breech had blown apart apparently because the gun had been underbreeched by the armourers in error. The first rounds entering the gun had been crushed and caught fire. We were fortunate that we had not been hit by some of the flying shrapnel, some of which was found in the rear of the cabin,

Our convoy work continued, interspersed with some night searchlight operations to give the batteries some practice. From time to time we did patrols well out to sea to report on icebergs coming down from the north which might endanger ships. Jaggs Lewis flew many of these patrols.

In the early fall I was suddenly posted to Ottawa. I was reluctant to leave the squadron and Jaggs didn't want me to go but the move was inevitable. Lin and I and Mike, our dog, plus our baggage, headed for Ottawa in the car. We got as far as Moncton before we were stopped by the RCMP and told to return to Dartmouth, where we found that the posting had been cancelled! We moved back into the quarters we had just vacated and I went back to my flying routine with the squadron, but not for long. A posting soon came through transferring me to No. 13

(Operational Training) Squadron at Patricia Bay, BC.

Once more, Lin, Mike and I crawled into the car to make a mad dash across Canada (we were not allowed to cross the US border as members of the armed forces) to arrive on the West Coast in quick time and with winter beginning to set in. I was actually supposed to go by train but I wanted to get Lin, our baggage, our dog and the car out to my new location, so driving seemed to be the answer.

21

Not So Merrily We Roll Along

Strangely enough, that drive to the West Coast was one of my most dangerous adventures. We packed the Dodge carefully with our gear and spread the dog's sleeping blanket over part of the back seat, Mike's own spot. Since we were not allowed to cross the border, I had planned our trip as an all-Canadian venture. There was no Trans-Canada Highway then, just a series of interconnecting gravelled roads, with a big gap around the north side of Lake Superior where no complete road existed. I found out from the shipping company that the last ship of the season carrying passengers and cars across Lake Superior was due to leave Sault Ste. Marie for Fort William in a few days. We would have to connect with it for sure or we would be completely out of luck.

Lin was a very good driver, so we planned to drive as long as possible each day and to spell each other off every 50 miles or so. We started from Halifax about October 25, 1940, hoping to average 50-55 mph, which on gravel roads was a good speed. On our first day we had a blow-out in New Brunswick but I put the spare tire on and we pushed on. I had the flat repaired while we were having a meal later on. There wasn't much traffic so we just drove on to the beat of stones thrown up by the wheels against the underside of the car. We stopped periodically to stretch our legs and walk Mike, and whenever we stopped for a meal we also picked up some food and water for him. He seemed to enjoy the trip. We made a long run that first day, pulling into Quebec City well after midnight, having been on the road about 16 hours. We managed to get a room for a few hours in the Chateau Frontenac, where they even let Mike into the room.

After a few hours sleep we started out again and, after another long day, managed to find a room for the night not far from Sault Ste. Marie. The next morning we arrived at the docks in time to have the car put aboard and to get a cabin. Poor Mike wasn't allowed to stay with us but was taken down into the bowels of the ship. The part of the trip we did in daylight hours was pleasant, although cold, but that night we ran into rough weather. The ship pitched and groaned and we had to hang on to our berths at times. Next morning we arrived in Fort William and made ready for the next leg of our journey.

From Fort William to Winnipeg there was a narrow, winding, gravel road through the bush. It rained and snowed alternately and the snow melted as it fell. After a long, rather slower drive we got to Winnipeg. We had been four days on the road. The following day we reached Swift Current, through rain and snow and over muddy roads. From there it was a comparatively short drive to Medicine Hat and Lethbridge, catching up with some much needed rest.

On day seven, we headed for the Rocky Mountains, through the Crows Nest Pass. We heard reports of heavy snow in the mountains so I tried, in vain, to buy a set of tire chains. As we pushed into the mountains we ran into more and more snow. Our big worry was to get over the highest part of the road, around the Rossland-Trail area. The road was a single gravel track with occasional wider spots for passing other cars, no guard rails, and at a higher elevation than it is today. We approached the highest part just after dark. It was snowing hard and getting slippery, which was awkward without chains. The police stopped us and suggested waiting until morning, but I was late for my reporting date at the squadron so decided to keep going.

We crept up along the highest part of the road in the driving snow. It was difficult to see the edge of the road in the headlights and there was a long drop below if we made a mistake. I was scared to use the brakes as, without chains, we could skid. We crossed the highest area and started down, keeping the car in lower gear. We were both rigid with fear but had no choice but to keep going. Fortunately we didn't meet a single car on the road that night. Everyone else has more sense than we did! Later we drove into Grand Forks, cold and tired, found a room and more or less collapsed, but we believed the worst was over. It wasn't.

The following morning we were off again. For some time everything seemed to be going well. We passed the southern end of the Okanagan Valley and were on our way to Princeton when, near Keremeos, we hit a sharp rock in the road which practically cut one of our front tires wide open. I put the old spare tire on again and in Keremeos found a new tire and installed it, but we had lost a lot of time. We went through Princeton and then headed north to Merritt and Spence's Bridge.

It was dark and snowing hard again as we approached Merritt. At one point in trying to follow the narrow winding road, Lin had the window open on her side and was trying to see through the driving snow. We could hear the roar of water somewhere nearby but couldn't see it. We were afraid we would go right off the road and into a river, but somehow we crept along. Eventually we groped our way into Spence's Bridge to find that the area was really snowed in. The hotel was filled with people who were waiting for the road to be cleared in the morning. We had to sleep in the lobby but at least it was warm.

The next morning, waiting for word that we could proceed, I scouted around, finally found a set of tire chains and had them installed. I would now be able to make some use of the brakes, if necessary. The road down the Fraser Canyon was rather narrow also and, since we were southbound, we found ourselves mostly on the outer edge of the road when we met other cars. Since the visibility still wasn't good, I was afraid to allow the cars on the inside any more width than was absolutely necessary. At one place a car suddenly appeared coming our way around a rocky corner. It had a large pair of antlers strapped to the front bumper. The antlers projected on our side. I gave as much room as I dared, but it wasn't enough. Our front fender neatly knocked the outer end of his prize antlers off. We didn't dare stop because of road conditions so we never knew what the driver thought of the matter. It wouldn't have been pleasant in any case.

After we passed Hope we left the snow behind, and I pulled off to remove the tire chains. We hadn't used them very much. Soon we were in Vancouver, where we took a night boat to Victoria, with the car below decks. The next morning after breakfast we drove from Victoria to Patricia Bay aerodrome, where I reported in. We were late but had actually crossed Canada in 10 days, without a modern Trans-Canada Highway, and for much of the trip in winter conditions. We were glad it was over.

22

West Coast Ops

At No. 13 (OT) Squadron headquarters at Patricia Bay everything was upset as the squadron had just moved from Vancouver and was not yet working smoothly again. The adjutant was F/O W.B. "Bill" Purvis. He and his family were to become our very good friends. He took me in to see the Commanding Officer, S/L R. "Dick" Briese with whom I had landed in the Stranraer on Sable Island about nine months earlier. He showed me around the unit and made me his second-in-command.

The squadron had two large flights, a seaplane flight commanded by F/L C.C. "Chuck" Austin and a land plane flight commanded by F/L H.J. "Harry" Winny. Both were experienced bush pilots, and old friends. The senior navigation officer was F/L J. Lorne Gray, a brilliant officer, later the president of Atomic Energy of Canada. The senior supply officer was F/L J.K. Lynch, otherwise known as "Flash." He had retired after long years in the US Navy and then come to Canada and joined the RCAF. In addition to supply duties Flash helped Lorne Gray and often accompanied navigation exercise flights out over the water. Another bush pilot in the unit was F/L A. "Archie" Vanhee. Archie acted as Chuck Austin's right hand man in the seaplane flight.

Dick Briese took me over to meet the Station Commander, W/C L.E. "Larry" Wray. I had met Larry previously in the Northwest Territories, where he showed up at times flying RCAF aircraft. All in all I was pleased to find so many old friends on one station. We moved into a furnished house just a few miles from the aerodrome on the gentle slope of a hill near John Dean Park and made ourselves comfortable.

The role of No. 13 Squadron was to train pilots in anti-submarine work, convoy patrols, bombing, machine guns, etc., so that they could take their place in coastal squadrons. In addition the squadron also acted as a strike force if and when needed, using Hudsons and the most experienced pilots in the unit. The landplane flight had Hudsons, Lockheed 10s and 12s and, later, Bolingbrokes and Northrop Deltas. The seaplane flight had the Grumman Goose, Norseman and Vedette, and later on some Stranraers. I got busy and had myself checked out on the Goose amphibian, Norseman floatplane and the Vedette flying boat so that I was fully qualified to fly any of the squadron's aircraft.

In November and December, 1940 and January, 1941, I was busy flying aircraft of one type or another on navigation exercises with some of the student pilots, practicing bombing and gunnery. On one of these flights I was accompanied by Rip Ripley (the ex-armament officer of 11 (BR) Squadron and now a squadron leader). Rip was now on strength at Western Air Command headquarters and visited our squadron regularly.

We learned at this time that my brother Dick, of TCA, had been loaned to the British Overseas Airways Corporation section of the Royal Air Force Ferry Command as chief inspector, to help achieve a high standard of maintenance on the Transatlantic operations. Dick did an outstanding job from 1940 to 1945 when he returned to TCA as superintendent of overhaul at Dorval, Quebec.

We also learned that my youngest brother, Eric, had also become involved in the war. He joined the RCAF in the accounts branch since he was a young banker in civilian life. He also returned to civil life in 1945-46. My sister Ida's husband, C.G. "Murph" Smith, was a civilian air engineer with TCA. He could not enlist because of a heart condition so he carried on as a civilian engineer until his death in 1961.

On January 29, 1941, I was asked to fly a Grumman Goose to Yorke Island, a military base at the northern end of the straits, to bring out a man very sick with spinal meningitis. I was accompanied, as second pilot, by our station medical officer, F/L Wally Brewster. It was a flight of nearly two hours each way but we finally got the sick man to hospital. We were a bit fidgety for a while wondering if we would catch meningitis or polio, but we didn't.

The work of qualifying operational pilots continued. We managed to get each group through and on their way before the next arrived. Every now and then we would retain an outstanding pilot and put him on staff. Such a one was P/O (later F/O) D.Z.T. Wood. Zack Wood was the son of the Commissioner of the RCMP. He did a good job and later was killed overseas in an airplane crash. Another good job was done by F/O Hal York. Yorkie had been a commercial pilot in the US, specializing in crop dusting. He joined the RCAF but after the US entered the war he transferred to the USAAF and was sent to Guadalcanal, where he was killed.

Around this time word was received that the Governor-General was going to visit the station. We polished and painted everything and spent hours at parade drill. We practised and practised in various formations from flights up to wings. The Governor-General arrived in an aircraft flown by my old friend Marlowe Kennedy. We went through our inspection and march-past, most of it in heavy rain, and when we broke off to go and meet the Governor-General in the Officer's Mess, we were sodden right through. The line of officers paying their respects were so wet they smelled like a Chinese laundry. We were pleased when it was all finished and could get back to our flying.

Lorne Gray had devised a mechanical navigation trainer, which was a tremendous effort. He not only planned and engineered the whole thing, but had the parts made. It was a great help in our navigation training and, I believe, was the beginning of synthetic navigation training in the RCAF.

In May 1941 Dick Briese was promoted to wing commander and I to squadron leader. Needless to say we were both pleased. During this month began the black-out checks over Victoria. When a black-out was called, one of us would take to the air, accompanied by Army observers, to fly around over the area looking for weak spots in the black-out.

Our staff was increased again by F/O F.E.W. Smith, one of our operational pupils. Frank Smith was a good hard-working pilot, who later was awarded an Air Force Cross, and who eventually went to India-Burma and won a Distinguished Flying Cross. Two more US pilots had joined us, F/L Bob Rizon (who won an Air Force Cross in the squadron for courage in fighting a fire in his Stranraer) and F/O Gale Schooling. Later the staff was further augmented by my old second-pilot-navigator from 11 Squadron, Bill Newson, now a flight lieutenant.

In early June Dick Briese, Lorne Gray and I flew to Rivers, Manitoba, and Winnipeg to visit navigation schools operating in the British Commonwealth Air Training Plan to see how their navigation training was done.

We had a party in the Mess the night before but were ready to go early the next morning in spite of sore heads. It invariably seemed that the severity of the weather was directly proportional to that of the hangover. We took off on June 8 in our Lockheed 12, number 7647 and headed east over the Rockies. The weather meant instrument flight. There was icing all the way across. We had no oxygen equipment in the Lockheed 12 so I was pleased to have had my TCA piloting experience across the Rockies. We landed at Fort Macleod where the Commanding Officer was W/C A.L. "Art" James (later Air Vice Marshal, now deceased). He gave us a drink for our hangovers and had us stay the night. The next day saw us at Rivers.

We spent the evening going over all the navigation training equipment and methods of training with the Commanding Officer and S/L F.R. Miller. Frank Miller had been at Camp Borden when the Mess fire had occurred and later had given me the Link Trainer course at Trenton while I was with TCA. By late evening Lorne Gray had all the information he needed and next morning we flew to Winnipeg to look over facilities there. On the 11th we flew back to Macleod and on the 12th, back to Patricia Bay, in much better weather this time.

In the meantime our Station Commander, W/C Wray, had been posted away and been replaced by W/C John L. Plant. John was an aggressive officer who liked to get things done quickly and do a lot of flying. We checked him out on most of the aircraft and he used them a lot. He and I often flew together. Dick Briese was posted overseas and I replaced him as Officer Commanding the squadron. Dick was eventually killed carrying out a reconnaissance flight along the coast of Norway.

In early September, we heard that we were going to be getting some of the new Bolingbrokes and also some Stranraers. The first one of each was to be picked up in eastern Canada immediately by our own squadron personnel. I decided to deliver the Bolingbroke myself and have Chuck Austin pick up the Stranraer. A party of airmen was selected to go as servicing crews and I picked Zack Wood to be my co-pilot navigator. Our complete party was to go east by train. By coincidence J.A.M. "Jack" Austin, president of Austin Airways, and his wife, Helen, were at Patricia Bay, visiting Chuck and arranged their return trip on the same train. We departed about September 15, leaving Harry Winny in command of the squadron.

Everybody in our party had a supply of liquor to last the trip, but we found that sitting watching the scenery go by, or talking about bush flying, caused the liquor to be pleasantly but more speedily used than planned. By Winnipeg we were all very dry. A carefully planned operation was organized there to have members of our party quickly take taxis and speed off to the liquor store while the train stopped. The liquor party barely had time to complete its mission before the train pulled out, but our vital supplies were successfully renewed.

To avoid a repeat of such a calamity, Jack Austin sent a telegram to his brother Bill in Chapleau, Ontario, and one to Austin Airways' chief engineer, Frank Russell, at Sudbury. As a result Bill met the train at Chapleau with a new supply and Frank met us in Sudbury with more. We were well taken care of. Jack and Helen Austin got off the train at Sudbury; and Chuck and his party at Ottawa, where his Stranraer awaited. The Bolingbroke party continued on to Montreal.

On September 20 we reported to the Fairchild plant at Longueil where the Canadian Bolingbrokes were made. The Bolingbroke was the Canadian version of the British long-nosed Blenheim bomber-reconnaissance aircraft. The test pilot at the Fairchild factory turned out to be my friend J.H. "Red" Lymburner, formerly of Canadian Airways, and one of my students on the Laird instrument flying course. Red showed me my aircraft then it was rolled out and we started up. Red went through the cockpit check with me; taxied out, took off and did a circuit. We changed places and I did a circuit and a respectable landing. Red was satisfied and declared me duly checked out.

Next day I took the Bolingbroke on a short familiarization flight with Zack Wood and my crewman aboard and in the afternoon we flew to Ottawa as I wanted to see how Chuck was coming along with the Stranraer. He was ready to go on the 23rd so he took off for Remi Lake, Kapuskasing. I headed for the Kapuskasing aerodrome.

As my aircraft was faster, I arrived well ahead of Chuck and was a bit concerned as the weather was deteriorating. Zack and I drove out to Remi Lake to meet Chuck who appeared and circled in the Stranraer just as a snowstorm blew up. Chuck made his approach in poor visibility, tried to land and bounced hard, then put on his engines again and disappeared in the snow. I was getting worried when he showed up again, felt his way down on the approach and landed safely. With the aid of boats we managed to get him moored to a buoy, with some difficulty as it was now snowing and blowing hard. The weather forced us to stay in Kapuskasing for five days. We could not do any instrument flying with the new aircraft, still being unsure of the instruments and radio.

On the 28th we took off for Sioux Lookout, where Chuck landed on the lake and I used the airstrip. After refuelling we took off again but here our routes separated. Chuck was to follow the more northerly route which was suitable for flying boat operations, while I flew west along the regular airways. Between Sioux Lookout and Winnipeg my starboard engine began intermittently cutting out. I couldn't pinpoint the cause and called Winnipeg to advise that I would need the engine looked over.

On arrival I parked off on the grass where a repair and refuelling crew met us. Three or four airmen approached, led by a flight sergeant whom I recognized as my old friend and air engineer Sammy Tomlinson. I had seen him off on the troopship for overseas in June 1940, but he was back in Canada and had been promoted. After a happy greeting, I told Sam about our engine trouble. He soon had his men checking things out.

Meanwhile, an RAF group captain on exchange duty in Canada joined us. He introduced himself and expressed interest in the new Bolingbroke. He asked about its approach speed. I gave him the airspeed figure I used, which I had arrived at by stalling the aircraft and then adding sufficient speed to give positive control under gusty or turbulent conditions. It was a safe, comfortable approach speed. The group captain told me that I should be approaching slower and that as soon as the aircraft was ready he would take me up to demonstrate what he meant. I was not happy at this prospect as I always liked to fly an aircraft as it suited me.

Sammy Tomlinson was listening in so I gave him a bit of a wink. He understood and made sure his men did not complete the engine repairs. The group captain eventually could wait no longer but promised the next time I landed in Winnipeg he would show me what he meant. We shook hands and he strolled away. As soon as he was out of sight Sam had the men put the cowlings on, as the engine problem had been corrected. I thanked him, then we took off. During the next two days we flew west in reasonable weather and returned to Patricia Bay. A day or two later Chuck Austin and the Stranraer arrived safely.

While I was away, W/C Plant had damaged our Lockheed 12 while landing. Our airfield was under construction in places, particularly along the runway edges. John had apparently veered a bit on landing and slid into the construction area. Though a good pilot, he had not been fast enough in correcting the aircraft's tendency to swing on the ground. He was a forthright officer who immediately officially criticized himself and signed his own pilot's log book, certifying that he had committed an error of judgement. John could be the only RCAF officer who has officially punished himself.

By now I was feeling at home in the Bolingbroke. It was a fine airplane to fly, very smooth on the controls, but it took a while to get used to its British air brakes. We added three or four more and managed to qualify most of our staff pilots on them. It was unfortunate that the plane later gave us a lot of trouble, mainly due to the failure of its hydraulic pump. There was only one pump on one of its two engines and it had a habit of failing at the same time that the emergency hand pump failed. The result was a number of belly landings when it was impossible to get the undercarriage down. Eventually the aircraft were sent away for modification.

In November we lost a Stranraer piloted by Sgt. Pilot Bliss and crew. It was being ferried from the east when it ran into bad weather and slammed into the side of a mountain near Howe Sound. We searched unsuccessfully for it. Years later, parts of it showed up in an avalanche.

On December 7, 1941, the Japanese attack on Pearl Harbour changed our operations overnight. The squadron was placed on Number One alert, all leaves were cancelled and for some time we were not allowed to leave the air station. Alert crews slept in hangars at night near the aircraft, which were fully armed and ready to go. The main strike component of the squadron was the Hudson unit. We had many reports of enemy submarines off the US or Canadian coasts, which usually turned out to be false but which had to be investigated by patrols, so that when not standing by ready to go, we were out over the Pacific patrolling. Our training had stopped temporarily. The station Commanding Officer, John Plant, had approved a plan for the ground defence of the station. All the units had to take part in this operation and practised it regularly. Our wives and families were moved to the mainland in case the Japanese fleet should swing our way. When it was later clear that this wasn't likely to happen, the families were moved back.

Our stand-by alert and patrols continued. On January 22, 1942, I was detailed to fly to McChord Field, Washington, as captain of a new USAAF Hudson Mark III. With second pilot, Lt. Itz, USAAF, John Plant had flown it to Patricia Bay from McChord. I was to fly it back to check out some of the American squadron commanders at McChord, as the Americans had only just taken delivery of them. I spent four days doing circuits and landings and so on with the Americans and managed to check them out satisfactorily. I returned to Patricia Bay on January 27.

Shortly after, John Plant was posted overseas to command No. 413 flying boat squadron in Ceylon. We had a great party in the Mess to wish him luck and then I flew him to Vancouver to catch an eastbound TCA flight. He flew a 413 Squadron Catalina from England out to Ceylon. He was replaced at Patricia Bay by G/C A.J. "Jimmy" Ashton.

Throughout April and early May 1942, we did a number of patrols out to sea and also reintroduced training. The Japanese had occupied Kiska in the Aleutian Islands and it appeared as if they might move along the Aleutian chain to Alaska. Senior RCAF and Army officers, were to fly to Alaska to discuss with the American commanders the possibility of sending Canadian squadrons there, and to determine which Alaskan airstrips they might use.

For this mission I was to be the captain of our Hudson 774 with F/L J.K. MacDonald as second pilot-navigator. We were to carry G/C A.H. Hull and Col. Mullaly, Canadian Army, and some of their staff. Weather along the coast to Alaska was particularly bad and there were few navigation aids, so I elected to fly over the Rockies to Lethbridge and Edmonton, then along the North-West Staging Route to Alaska.

We completed the first part of the flight as far as Edmonton by May 21, where we were delayed by weather and hydraulic trouble. By the night of the 23rd we had arrived at Fort St. John, where the weather again held us up. On the 25th we landed at Whitehorse, Yukon, and on the 26th at Ladd Field at Fairbanks, Alaska. Two days were spent here in talks with the Americans. It was interesting to see how modern Fairbanks compared with many of our Canadian bush settlements. One thing in its favour, of course, was its comparative proximity to the Pacific Coast.

On May 28 we took off for Anchorage, at the base of the Aleutian chain. The poor weather meant instrument flying. We flew very high and apparently got a bit off course, because in passing through some breaks in the cloud we suddenly saw the rugged face of Mount McKinley dead ahead. A quick change of course straightened things out but it gave us quite a surprise. We landed at Elmendorf Field, Anchorage, and found everything on a busy, war footing. Everyone, troops and civilians alike, believed the Japanese would push right up the Aleutian chain, certainly as far as Anchorage. A comprehensive station defence plan was in operation, even to some slit trenches around the base. After being installed in our quarters in one of the barrack blocks, we found that, as a crew, we were free for the evening. The senior officers were involved in discussions with the Americans but we were not needed. We decided to go into Anchorage and have a look around.

We took a cab downtown where in our RCAF uniforms we were very welcomed, as everyone was concerned about the Japanese threat. We soon found ourselves a pleasant-looking bar. The bartender enquired about our uniforms. We told him we belonged to the Royal Canadian Air Force and soon all the patrons were buying us drinks. Having consumed a fair quantity of alcohol, we decided to look for a place for dinner. Along the street people greeted us and one fellow insisted we have a drink from a bottle he had in a paper bag in his pocket! We found a good restaurant and again were overwhelmed by the Alaskan hospitality.

After dinner we took a cab back to the air base—we had all the food and drink we could take, but wished we had stayed in Anchorage. Apparently a Japanese night reconnaissance flight had been reported coming along the Aleutians. The camp was blacked out and all those not needed for base defence were ordered to the slit trenches. We objected but didn't get very far. We crouched in the trenches for a while until the alert was called off, then were allowed to go to our quarters, our teeth chattering with cold. Next day we flew back to Fairbanks, where our main mission was completed. G/C Hull had agreed with the Americans about which airfields our forces would use and what squadrons would be sent. In passing, it had been suggested that I might be offered a post on the staff of the US commander in Alaska, Gen. Buckner. I indicated I would be pleased at this if it could be arranged with the RCAF.

We left Fairbanks on May 30, landing en route at Whitehorse, Fort St. John, Edmonton, Lethbridge and Vancouver, reaching Patricia Bay on the night of June 1. As a result of this mission, one bomber reconnaissance and two fighter squadrons were sent to Alaska to support the Americans.

At Patricia Bay we found things a bit tense as there was fear of a large Japanese fleet heading our way. It was stopped later by the Americans near Midway but, before that action took place, a Number One alert went into effect on the West Coast. Again we were confined to the station, ready for action.

We stood to for a number of days without anything happening, which resulted in a drop in morale. I suggested to the Station Commander that it might be a good thing to allow our wives to visit the next Saturday evening in the Messes, where we could at least see and talk with them. He agreed and everything was arranged. At this time half our station was used by a Royal Air Force operational training unit flying Hampdens. This OTU was not to conduct actual operations from our base.

The night of the gathering in the Messes, June 20-21, while talking to Lin, I noticed an airman come in with a message form which he took to the the RAF commander. Upon reading the message, he immediately went into a huddle with some of his senior officers after which they left the Mess hurriedly. I went over to our own Commanding Officer, G/C Ashton, and reported this. He spoke to the RAF commander and then called me—an enemy submarine was shelling Estevan Point on the west coast of Vancouver Island and our strike force was to take off immediately. I grabbed my top strike force pilots and, after calling our hangar to have the aircraft readied, we sped off to the flight line. We could hear engines running in the darkness on the RAF side of the field. It was clear that they had decided to go after the submarine themselves and they had a head start as a result of the message going to the wrong people. I rushed my crews into action, and then led them out in my own aircraft. As we taxied for take-off I saw that something had gone wrong and the tower radioed that the RAF had managed to block both runways.

One Hampden had collapsed its undercarriage on the main runway and another had run off the edge of the second runway and was blocking it. We were stuck, as between the two runways much construction work was going on. The tower advised that take-offs were impossible and that the enemy submarine had stopped shelling. Back at the hangars I have never seen such a group of angry people. We were so mad we were actually in tears. Later, apparently, the RAF apologised for the trouble, but the damage had been done. It was too bad because our Hudson strike force was composed of our very best crews, well trained, who could have got to Estevan Point in the darkness and had a chance to bomb the Japanese submarine. We returned to the Mess but our evening had been totally ruined.

As a result of the US victory at Midway, we were stood down from our alert and normal operations resumed. I received instructions to fly a group of senior officers to San Francisco for an urgent conference with the Americans over West Coast operations. I took our old Lockheed 12 so that all the Hudsons would be available in case of trouble while we were away. The passengers were to be A/V/M Leigh Stevenson, the Air Officer Commanding Western Air Command, Lieut.-Gen. Stuart of the Army, and Cmdr. Beech of the Navy. The crew was my crewman Cpl J. Romanow (later Colonel J. Romanow, RCAF Regular Force) and myself.

We took off on the morning of June 23, refuelled at Portland, Oregon, and thereafter ran into troubles. As we climbed over the mountainous route, our fuel pumps started to give trouble, the red fuel warning lights coming on intermittently. We had to do a lot of pumping with the emergency hand pump to keep the fuel pressures up. We landed at Eugene, Oregon, and Cpl Romanow checked things over. There were no fuel leaks so the trouble was with the pumps themselves. We got going again but had to land at Red Bluff, California, and rest again. Finally we reached Hamilton Field, across the Bay from San Francisco, where our passengers, who had had a nerve-wracking ride so far, departed for their conference.

The USAAF engineers pitched in and the pumps were taken apart. Apparently long wear had resulted in the clearances in the pumps being too great. All nearby bases were tried for replacements, but none of the right parts were available. The old ones were re-installed for our return. I did a test flight from Hamilton Field but the red lights glowed again so we knew we would have more troubles. The passengers were not too pleased about the prospect of flying with bad pumps, but we couldn't help it.

We took off on the 27th and although we had some fuel pressure, Joe Romanow and I had to pump our way to Portland where we landed when the weather deteriorated. Next day we pumped our way to Patricia Bay. The passengers and crew were relieved to get back to base and my arm was tired for some time.

On return I heard that I had an immediate posting from Patricia Bay. I was sure that this meant Alaska and I was very pleased, but it turned out to be a posting to Air Force Headquarters, in Ottawa. This angered me and I spoke to A/V/M Stevenson. He tried in vain to get the posting changed. Thus, I turned over command of the squadron to Chuck Austin, who was shortly promoted to squadron leader. Lin and I once again packed our gear, and Mike, the dog, into our old Dodge and made ready to drive to Ottawa.

23

Foundations of a Command

We drove east through the United States. Since the US was now at war, Service people could now cross the border without risking internment. We had a pleasant but hurried trip, coming back into Canada through Detroit and Windsor.

We took a nice furnished house in Ottawa on Glen Avenue. I was to be a member of the Air Staff at Headquarters with particular reference to air transport and the ultimate object of building up an Air Transport Command. The only real air transport unit existing in the RCAF was No. 12 (Communications) Squadron at Rockcliffe Air Station, commanded by S/L H. Marlowe Kennedy. He had managed leave from TCA shortly after I had left the airline.

At Headquarters there were many familiar faces. The Air Member for Air Staff was A/V/M N.R. Anderson, formerly of Eastern Air Command. There was A/C K.M. Guthrie, later A/V/M; W/Cs C.L. Annis and L.E. Wray, later A/M and A/V/M respectively, and others.

One of my operations at this time was to send a flying boat to Greenland to land near Godthaab, in spite of very heavy ice conditions, and bring out the senior Canadian diplomat who was urgently needed in Ottawa. I was able to get hold of S/L Jack Hone, an experienced bush pilot in civil life, and a suitable crew who completed this difficult mission. Jack was awarded a Bar to his Air Force Cross in 1945 for a number of such special flights.

We also planned and organized a service flight which No. 12 operated on a schedule between Ottawa, Montreal, Moncton and Dartmouth, using Lockheed 10As. I flew a few trips as a check pilot to keep my flying up. The service was nicknamed "The Blueberry." No. 12 also carried out all VIP flights from the Ottawa area. All pilots were trained to airline standards by Marlowe Kennedy and his staff, something absolutely necessary in safely carrying the high-level brass. In addition, Marlowe trained some pilots to carry out VIP seaplane and flying boat operations.

On one "Blueberry" flight, Marlowe and I operated as the crew, alternating as pilot and co-pilot. Flying near Moncton at night and in heavy overcast, both engines suddenly stopped. We both reacted immediately as captains and reached for throttle and fuel controls. We locked hands in the darkened cockpit reaching for the same controls. After we got our fingers untangled and reset the fuel controls, the engines came back to life. We laughed about the incident afterwards but it was a good case of too many captains in one cockpit.

Towards the end of 1942 pressure mounted to get a military aerodrome into operation at remote Goose Bay, Labrador. Vast amounts of material had to be flown in using a temporary strip. For this airlift we organized No. 164 Transport Squadron to operate from Moncton under command of another former bush and airline pilot, S/L Bruce Middleton. Bruce and his men did a terrific job with their Dakotas and Lodestars, keeping the freight moving into Goose Bay in tough winter flying conditions.

No. 124 (Ferry) Squadron, which had existed for some time and was commanded by S/L E.O.W. "Ernie" Hall, another experienced pilot, was brought into the Transport scope. It was busy ferrying large numbers of aircraft from the factories and overhaul bases to training schools and operational squadrons across Canada. Dispatching control in a squadron of this type was vital and was presided over by F/L (later S/L) Jack Carson, a former Canadian Airways traffic dispatcher.

At this time arrangements were made for Marlowe Kennedy to travel to London to study RAF Transport Command operations. He flew over in the bomb bay of a Ferry Command Liberator, returning a month or two later with much valuable information for us.

In spite of the growing air transport complex requiring a greater centralized control, I managed to fly enough to keep my standards up. An efficient headquarters staff, including transport traffic experts like F/Ls Leslie Collins and Dick Forrest, made this possible. I did many trips around the country to see personally what was going on and what was really required.

Early in 1943 it became evident that a new transport squadron was needed in the west especially to cover the new North-West Staging Route to the Yukon and Alaska. I took S/L Humphrey Madden from his position as second-in-command to Bruce Middleton and sent him to Edmonton with a few Dakotas and personnel taken from varioud units. At first they had to operate under canvas. It took some time to get the new squadron approved but the Chief of the Air Staff, A/M Lloyd Breadner, got the unit approved as No. 165 (Transport) Squadron.

Breadner had also tried to get approval for a new command, RCAF Air Transport Command, but Cabinet was not ready to approve another command so a new proposition was tried. The Directorate of Air

Transport Command was created at Air Force Headquarters and I was promoted to wing commander as its first Director. The new Directorate operated exactly as a Command headquarters would, even though it was within Air Force Headquarters. We commanded and controlled the operations of units in the field directly from the Directorate. I created three wings within the Directorate, the Domestic Transport Wing, the Overseas Transport Wing and the Ferry Wing. A qualified officer and staff controlled each wing from our headquarters. The need for a Ferry Squadron located in western Canada was pressing, so we also created No. 170 (Ferry) Squadron at Winnipeg under control of the Ferry Wing of the Directorate.

Throughout 1943 we were busy keeping everything operating smoothly in spite of continually growing commitments. In August I flew a new Dakota to Edmonton and then with "Hump" Madden did a tour of the North-West Staging Route in one of our Lodestars. I wanted to study the facilities for air navigation along the West Coast, so we flew from Whitehorse over the White Pass and then down to Annette Island, Alaska, where an RCAF unit was located. At lunch there I met S/L Gordon Diamond. He agreed to join our organization and on return to Ottawa I managed to have him transferred to us. This move was a good one for air transport. I had W/C Marlowe Kennedy moved up to take command of the Domestic Transport Wing; and Gordon Diamond, promoted to wing commander, to command No. 12 Squadron. W/C Ernie Hall was moved up from 124 (Ferry) Squadron to command of the Ferry Wing.

In the middle of this busy period I went to Orlando, Florida, where the USAAF was conducting special instrument approach and blind landing tests. I was to fly with the Americans and report on the operations. We flew in B-25 aircraft, a number of which were badly damaged in the landings. I was pleased with some of the instrument approach equipment but felt there was much to learn about the actual blind landings.

On my return I found that the Minister of National Defence, Col. Ralston, was pushing for an overseas air transport service. He had just returned from the United Kingdom and Italy, where he had found that the troops' morale was being affected by poor mail service from home. The Minister for Air, C.G. Power, decided that a service should be started as soon as possible to carry troops' mail and other priority items from Canada to the United Kingdom, then along the North African coast to Cairo, and thence up into Italy. A/M Breadner had gone to England as Air Officer Commanding-in-Chief at Headquarters in London and had been replaced by A/M Robert Leckie as Chief of the Air Staff in Ottawa. The Air Marshal set a target date of Christmas 1943 to have the first regular air mail delivered to the troops in Italy. We only had about two months to plan and build the operation. In the end we didn't quite get the mail to the troops in Italy for Christmas, but did make it by December 29, despite the toughness of the assignment.

The only four-engined aircraft available at the time were six Boeing B-17 Flying Fortresses being used for training by the USAAF in Texas. They were immediately purchased and ferried by American pilots to Rockcliffe. Just before their arrival, we created a new squadron, No. 168 (Heavy Transport), for the overseas job. I moved W/C Middleton from 164 Squadron to command No. 168, replacing him with S/L (later W/C) Charles Hoyt, who also did a great job.

We began getting tour-expired bomber pilots to fill our transport requirements. This was ideal, as the pilots were accustomed to heavy aircraft. We then retrained them to transport standards.

Bruce Middleton soon got things moving with his new squadron. There was little time for a thorough check-out on the Flying Fortresses. This was difficult for the ground engineers, who had to learn about the aircraft while flying was going full out. They also had to remove all armament equipment and install long-range fuselage tanks in some of the aircraft. A few later-model aircraft had extra wing tanks, nicknamed "Tokyo" tanks. Long-range cruising flights were done to various points over Canada following great circle courses in order to get the pilot-navigator teams working together and to train pilot-flight engineer teams in fuel conservation to get maximum range. It was tiring work, but we were soon ready for our first flight overseas. I would go along to see how everything worked out and to meet with A/M Breadner and A/C/M Sir Frederick Bowhill in London to arrange the use of bases we required in the Mediterranean area and Italy itself. We planned two flights, one following the other immediately because of the initial heavy mail loads. Thereafter we would follow a fixed schedule.

Our departure from Rockcliffe was set for December 15, 1943. Bruce Middleton was to captain the first flight with S/L B.G. Smith as co-pilot, and myself spelling off as needed. F/O Jimmie Irvine was the engineer, F/O La Brish the navigator (he was killed later on) and the wireless operator was F/O Dickson. The second flight was captained by F/L C.R. "Ronny" Knowles. Marlowe Kennedy was to act as Director in Ottawa during my absence. A/V/M A.L. "Art" James, then the Air Member for Aeronautical Engineering, thought we were moving a bit too fast and that the engineers had not had time to ensure that the aircraft were in good shape. It turned out that he was right, but this didn't become evident until we reached mid-Atlantic.

On December 15, 1943, our Fortress 9204 was ready for departure, mail load aboard and fuel weight arranged to allow for a landing at Dorval where more mail awaited. We were given a send-off by the Minister for Air, C.G. Power; the Deputy Postmaster-General, Mr.

Collican, Chief of the Air Staff, A/M Leckie; A/V/M K.G. Nairn; W/C H.M. Kennedy; W/C Doug Findlay; W/C C.G. Diamond and others. Bruce Middleton made a smooth take-off and 45 minutes later we landed at Dorval. Here unfortunately we ran into mechanical and weather trouble, delaying departure until the 17th, when we left for Gander.

The flight to Gander took five hours and gave us a taste of what was ahead. It was cold in the aircraft at higher altitudes in spite of the heaters. It was particularly cold in the rear where the .50 calibre gun mounts had been. Each gun mount had a sliding door which leaked cold air even when closed. We spent our time here when not busy up front. To reach the rear compartment we had to pass through the big bomb bays on a metal catwalk. The bomb-bay doors were held closed hydraulically and the mail load was stacked mostly there in the bays. The aircraft and engines had functioned well enough but we had not been able to test the Tokyo tanks in each wingtip of Fortresses from the F model onwards. For the test we had to run the main tanks fairly low before the Tokyo tanks would start their gravity flow. Our five-hour flight to Gander was insufficient for this, but in any case we were not expecting trouble.

At Gander we, along with many other aircraft being ferried overseas, were grounded by bad weather over the Atlantic. On the night of the 20th things looked a bit better so a late briefing was called for all aircraft leaving that night. Ronny Knowles' Fortress had caught up with us by now and also was due to leave that night. In the briefing room I was startled to see the familiar face of engineer Lew Parmenter. Lew had been Walter E. Gilbert's engineer in the north during our bush days while flying from Fort McMurray. It turned out that Lew had decided he could help the war effort best by flying with the young, sometimes inexperienced ferry pilots on the Atlantic. The weather briefing section was headed by Patrick McTaggart-Cowan, whom I was to see again many times.

We taxied out with a steady line of aircraft and took off, climbed into the darkness and the cold and set course for Prestwick, Scotland. We gradually climbed to keep above the weather until we were cruising at 27,000 feet with a good tail wind. All of us were then up front and all using oxygen. We were past the point of no return (where the remaining fuel would only permit continuing to our destination) and something over seven hours out of Gander when Bruce reported, "I think something is wrong with the Tokyo tanks. They're turned on but the main tanks are still down."

Careful checks showed that the fuel was definitely not flowing from the Tokyo tanks into the main tanks. Bruce and I concluded that the valves operating the Tokyos must be frozen up. We knew the weather below with heavy icing, was not good but felt that we must get the Tokyos working. So Bruce descended into the cloud to get into warmer air. We kept on going down, hoping that the tanks would work. They didn't. Finally in the early dawn we were under the cloud ceiling, just above the water. The wind was blowing a gale and the seas were running high. We knew that if we had to ditch, our chances of survival were small. The Tokyos still wouldn't work and now we couldn't afford fuel to climb back to altitude. Our rough calculations showed that we probably wouldn't have enough fuel to reach the west coast of Ireland.

At this point we sent out a mayday signal, the international distress signal which is only sent in dire circumstances. It was acknowledged, but there was little or nothing that could be done to help us. The engines were leaned out as far as possible. The weather ahead was bad but we were advised to head for an emergency airstrip at a place called St. Angelo, just in from the Irish coast. Rockets would be fired from the strip to guide us if we got near it.

We were all tensely waiting to see what would happen, as we were not sure of the accuracy of our fuel gauges. One thing was clear — the main tanks were not receiving any fuel from the Tokyo tanks. Everyone concentrated quietly on his job but wondered when the engines would quit. A number of silent prayers were undoubtedly being said, asking that the engines keep going. Finally the fuel gauges were showing empty and just then through the fog and rain we saw rockets bursting ahead. Bruce expertly lined up with the runway and landed safely. As the Fortress rolled to a stop, two of the engines died from lack of fuel. It was as close as that. We were met by W/C J.R. "Joe" Frizzle of the RCAF, stationed here flying anti-submarine patrols.

The weather was hopeless as far as getting away again was concerned. Joe Frizzle made arrangements to have our main tanks refuelled. We found out that the Tokyo tanks were still full of fuel, but would have to wait until we arrived in Prestwick to find out what the trouble was, and in any case the main tanks would take care of our final leg. Joe drove us to Castle Archdale, the area headquarters, where we met the area commander, G/C Martin "Costy" Costello, RCAF. He welcomed us with food, quarters and some much welcomed Irish whickey, and ensured that messages were sent advising of our safe arrival. We learned that Ronny Knowles had already landed in Prestwick. His fuel system had given no trouble, so he had remained at high altitude and carried on.

After a good night's rest and a relaxed visit with Costy and Joe we took off the next morning for Prestwick, where we were met by G/C Ted Underwood of the Canadian Postal Corps. He was in charge of all overseas Canadian mail and was to travel with us to the Middle East. Our total flying time from Gander was 9 hours and 55 minutes.

At Prestwick, G/C McIntyre of Scottish Aviation Company agreed to handle any maintenance that was beyond the capabilities of our flight engineers. He was also to troubleshoot our fuel system and make the necessary repairs. Ronny Knowles and his crew were to remain in Prestwick for the next flight, and we would take his aircraft, Fortress 9203, on to the Mediterranean while ours was being repaired. The next step was to go to London and see about bases, fuel, servicing, etc. throughout the Mediterranean area. Ted Underwood, Bruce and I climbed aboard a train at Kilmarnock and went down to London.

December 22 and 23 were spent in London in talks with A/M Breadner at his headquarters in Lincolns' Inn Fields and with G/C Geoff McDougall and A/C/M Sir Frederick Bowhill, both of RAF Transport Command. As a result we arranged to use bases at Gibraltar, Rabat-Sale in French Morroco, Algiers, Tunis, Cairo, and Naples and Foggia in Italy; and were to be supplied with fuel, food and accommodation, weather briefings, and anything else which we needed and was available. The accounting would be done between the two headquarters. Messages were sent to the commanding officers of all bases concerned, advising them of the arrangements.

During our stay in London there was some bombing by the Germans, mostly in the East End. We witnessed the terrible result of the raids. The smell of fire and rubble permeated everything; but the courage of those working and living here made a profound impression on us. We took a night train back to Prestwick, arriving on Christmas Eve morning. Weather checks showed that we could get away later that night.

We had our aircrew meal and briefing, and took off after midnight. It was now December 25. We had to fly west for a while before heading south in order to keep clear of the Bay of Biscay area where German night interceptors were busy. Daylight came as we were west of Gibraltar. After 9 hours and 50 minutes we landed at Rabat on a steel mat runway laid over the sand. The RAF station commander had his orders from London regarding us but wasn't very cooperative. The huts there for our RCAF personnel were in bad shape. I instructed the station commander they would have to be cleaned and spruced up by the time we got back from Italy. Even his Officers' Mess was in an unsatisfactory condition. After a subsequent meeting in London with Sir Frederick Bowhill, the commander at Rabat was replaced by a more suitable man. Everywhere else we received full cooperation.

We slept part of the day, having been awake all the previous night; in mid-afternoon we walked into Rabat's Kasbah, the old walled section of the city. We were advised to be armed and to be back before nightfall. Troops caught in the Kasbah after dark, both here and in Algiers, were sometimes killed and robbed of their wallets and shoes. We carried our revolvers inside of our battle dress instead of in a holster, to make it harder for them to be stolen. Ted Underwood, Bruce, B.G. Smith and I walked through the narrow, dirty streets of the Kasbah, watched people shopping at small, fly-infested markets and felt very sorry for them. We started back as dusk settled, but soon things looked a bit different. We thought we might be lost but didn't dare ask for directions as we had been told that such directions would deliberately be wrong. On we went, trying to look calm, until finally we saw the gate where we had entered, and were soon back in camp, a relieved little group!

December 26 we took off for Algiers in the morning. In the North African and Italian war zones, we were supposed to fly transports only by day, and then under 3000 feet; otherwise we were in danger of being shot at. After five hours of flight we landed at Maison Blanche aerodrome near Algiers. Here at a dinner meeting with the RAF we completed all arrangements for our air service through Algiers. On the 27th the weather was poor, so we had a look around Algiers.

Next day we flew to El Aouina airport at Tunis in 2 hours and 50 minutes. Here we were met by G/C George Ault, the senior RCAF administrative officer in the area. George had arranged for our refuelling and servicing and we took off shortly, landing two hours later in Catania, Sicily. Everything in the area was in a terrible state, as a major battle had been recently fought around here. The hangars were burnt and the wrecks of aircraft, German, Italian and ours, were just pushed off the runways out of the way. It was nearly dark so we tied our aircraft down and were driven to a Canadian Military Hospital, the only accommodation available for the night. It had been a German Military Hospital and had merely changed hands. We were given beds in one of the wards and then taken over to the Officers' Mess for a drink and something to eat. We met Col. Magnus Spence, the Commanding Officer, and some of his officers, including Capt. Frank Christie of Lethbridge. Col. Spence returned to Toronto after the war and we met again in 1954-55 when he became Lin's doctor. Frank Christie returned to Lethbridge and became one of the top surgeons in that area, and I had the pleasure of playing golf with him in later years.

We had a grand evening in the Mess, drinking captured German cognac, but as we had to get started early we went back to the hospital early. We had a good look at the flames from the active volcano Mount Etna before we went to bed.

Next morning Bruce and I, in a room with two beds in it, were awakened by two very large Sicilian women pushing hospital wagons with bowls of steaming hot water. They flung open the black-out curtains and let an icy wind blow into the room, then started to remove the bedclothes to give us a complete washing. We tried to tell them we were guests, not patients, but they didn't understand. Finally there was quite an uproar and a Canadian Army nursing sister appeared. She stood in the doorway, bursting with laughter, as Bruce and I tried to hang on to our blankets. After letting the battle rage for a while she told the Sicilian women to leave us alone. We could hear her laughing as she disappeared down the corridor.

Before long we took off for Foggia where we had to be careful as the front lines were not too far north of the aerodrome. We had been briefed to fly through a certain mountain pass, remaining below 3000 feet, and on breaking out onto the plains of Foggia to keep a sharp lookout for enemy aircraft. Our aircraft, although it looked formidable, was unarmed. We flew through the pass, but on entering the plain found there was a low-level air battle taking place between German and Allied fighters. They were streaking around near the mouth of the pass. Bruce opened the throttles and we got out of the way as fast as possible. As we neared Foggia, the tower told us to get on the ground quickly due to enemy activity. Two aircraft ahead of us were also getting down in a hurry, an American B-26 and just below it a Savoia-Marchetti three-engined aircraft belonging to those Italians who were now on our side.

Just ahead of us on the approach we saw the American B-26 land on top of the Savoia-Marchetti. The American pilot apparently didn't see it. The Savoia was driven into the ground at the end of the runway, where it exploded, and the B-26 skidded on its belly across the muddy field. Bruce flew down on to the runway through the flames of the burning Savoia. A number of Italians perished in the accident.

We unloaded some of our mail here for the front-line troops. It was December 29, making us four days late in our effort to arrive on Christmas Day. Arrangements were made for the regular service to follow, then we took off for an aerodrome near Naples, where we landed 30 minutes later and the rest of our mail was unloaded. Getting out of the aircraft, I hurt my back jumping from the rear door, flight bag in hand. My back straightened out again but did give me some trouble later on.

We drove into Naples in a military vehicle to a commandeered Italian villa set in ornate grounds enclosed by stone walls. It was a tranquil place, although some of the windows on the harbour side had been blown out. As there was no electric power or heat, it was cold inside. An Italian man-and-wife team looked after the place which, we were told, had belonged to a high-ranking Fascist who owned the Naples electric company. We slept on cots and rolled ourselves up in army blankets. Next morning I found my back had been pretty well chewed by some sort of bug.

We spent the day looking around Naples and ensuring that our requirements for accommodation and servicing of the airplanes and crews were in order. We visited Ted Underwood's army post office, a beautiful old theatre in central Naples, now piled full of mail bags. We also saw the dock areas which had been thoroughly bombed by our side, but with careful aim so that not much of the city was damaged.

December 31 we were loaded with return mail, flew over to Foggia again, which was now quiet, and picked up more mail. We flew back across to Tunis in 3 hours and 15 minutes. New Year's Day 1944, we were grounded with urgent maintenance work on the aircraft, so George Ault drove us out to the ruins of old Carthage, along the coast from Tunis. The car was a beautiful Alfa-Romeo commandeered from someone who had supported the enemy side. George told us that his headquarters was shortly to be moved to Naples — good news, as he could then help with our operation at its most important terminal.

On January 2 we took off with a load of mail for Cairo. We flew as far as Tripoli, where we wanted to look the facilities over to see if we could use them. In the air I began to feel sick; my throat was closing up and diarrhea and cramps were developing. We found the airfield in good shape except for burned hangars. Accommodation was excellent, having been built in pre-war days by Gen. Balboa, the Italian aviator. An RAF doctor who had spent much time in the Middle East was in the Mess. He said my trouble was "Gyppy Tummy," a common but distressing complaint in the East. He made me take some awful tasting medicine, then sat down with me while I drank a large stein of beer. We had been briefed in the United Kingdom not to touch alcohol if we became sick in our travels, but this doctor insisted on my drinking the beer. He said it would work and it did. Instead of spending the next two or three days lying in my quarters, I felt a lot better the next morning, so we took off in the morning for Cairo.

At times our route was near the coast, other times over the desert. We could see the tracks of tanks and other vehicles here and there where battles had been fought. Burned tanks and aircraft were lying around in places. After 6 hours and 30 minutes, we landed at Cairo West aerodrome, well out from Cairo and on the edge of the desert.

We went into Cairo in a light military truck with a canvas awning over the open body. We were met in Cairo by W/C Patterson, who commanded the RCAF administrative unit there and did a good job arranging for our regular transport service in that area. When we arrived the bags of one of our boys had been lifted from the back of the truck somewhere between the aerodrome and the city. Apparently this commonly occurred at places along the road where the trucks slowed down. We were accommodated at Shepheards Hotel but weren't impressed by it so moved over to Mena House, near the Nile and close to the pyramids and the Sphinx. Churchill and Roosevelt had held a meeting there shortly before.

After meetings at Heliopolis with the RAF area commander to confirm arrangements, we had a free afternoon before departing with another load of mail. We decided to explore the Great Pyramid of Cheops and the Temple of the Sphinx. Mena House arranged a dragoman, or guide, who turned out to be a pleasant, educated man whose family history involved guiding people through the pyramids. We went through the Temple of the Sphinx with its great roof slabs over the quarters of the high priests. Then we climbed part of the outside of the pyramid, through an opening and inside. It was dark and we hung on to each other in a line. The dragoman would light a strip of magnesium now and then and show us the pitfalls which were designed to waylay robbers. It was a bit frightening at times, particularly when we had to crawl through a narrow tunnel in pitch darkness to get into the sarcophagus chambers of Cheops and his lady. Now and then the blocks of the pyramid would creak and groan. We were impressed, but also a bit relieved when it was over and we were back in the sunlight.

We took off from Cairo on January 6 for Luqa aerodrome on Malta. Here we would refuel, then fly back to Italy again. We were leaving everything arranged behind us for our regular service and also cleaning up piles of accumulated troops' mail so that from there on the service would have a good start.

We covered most of the Mediterranean with no problems until we neared Malta. An air raid had been going on before we approached and as we ducked in and out of the clouds we flew into a large clear area with a German Ju 88 coming towards us. He must have been out of ammunition and we were unarmed. He ducked into the clouds in one direction and we did the same in another. Fortunately, we never saw each other again. Our flying time was 7 hours and 30 minutes as we landed at Luqa, where they were more or less permanently on alert. We remained overnight because of the lateness of the hour. We were interested in the soft stone houses of Malta; the stone is cut into blocks with big hand saws and most of the windows had no glass. Our cots had sheets which were shiny black with grease. The island had been raided so often that mechanics slid in and out of bed without time to take off their clothes. No one complained because it wouldn't have done any good. Food was not too plentiful either.

Valetta, the capital city, had been severely bombed, and everything seemed to be damaged. We went into an upstairs bar to have a drink. Buildings on either side had been bombed out, and the one we were in would sway in a strong wind, and have to be cleared for safety.

We took off next morning for Italy and flew through the mountains to Foggia in 3 hours and 15 minutes. After making mail transfers we flew to Naples, where we again changed mail bags. From Naples this time we were on the way home with all our pioneer work done. The next aircraft had arrived in the area with its mail, indicating that our service was beginning to operate behind us. One hour and 35 minutes later we had to go into Catania, Sicily, because of weather. It cleared again and we left for Algiers, landing after 4 hours and 10 minutes. We had flown past a huge formation of ships heading east up the Mediterranean. They had balloons flying from many of the ships and were obviously ready for action. We heard later that this was the formation carrying in the troops and supplies for the beach head at Anzio, north of Naples.

On arrival at Maison Blanche I was given two messages forwarded by Marlowe Kennedy in Ottawa. One said that I had been awarded the Order of the British Empire, Military Division, in the New Year's Honours List, and the other that I had been promoted to group captain. We stayed in the Aletti Hotel in Algiers that night, where we had a few drinks to celebrate The Aletti came under the control of the Town-Major of the Allied Forces in the area, so our accommodations didn't cost us much.

We left for Gibraltar next morning, arriving after 3 hours and 20 minutes. It was nearly always a challenge to land at Gibraltar, the landing strip being close to the Rock and having water at both ends. Turbulence from the Rock was usually felt during the touchdown. In addition, I was coming down with another touch of "Gyppy Tummy" so reported to the senior RAF medical officer. He confirmed my diagnosis, gave me some pills and told me to rest in my quarters for a couple of days. Thinking of the MO at Tripoli, I asked if I could drink some beer, but this MO wouldn't let me touch a drop. I took some of the pills, then went to the Mess for a beer anyway before retiring. The next morning I felt much better.

At this time a Hudson arrived from the Mediterranean, carrying the Air Officer Commanding-in-Chief of the RCAF Overseas, A/M Breadner. I was able to give him a detailed report of our accomplishments so far. He also was flying back, but to southern England so he got away just ahead of us. We left on the 12th and ploughed through weather right to Prestwick, with a flying time of 8 hours and 45 minutes. There I saw that our overseas service was now in full operation, with aircraft operating in both directions.

At Prestwick we were shown pictures of what had caused out trouble with the Tokyo tanks. The fuel could not possibly have flowed from the tanks as the fuel lines between the Tokyos and the main tanks had been cut and plugged or crimped off! This had apparently been done by the USAAF when they used the aircraft for training before we bought them. Apparently they had had trouble with air locks in the system so they had cut off the Tokyo tanks but neglected to enter this in the log books. Our engineers had not had time to spot the problem. The engineers at Scottish Aviation had by now got 9204 ready for us to take back across the Atlantic.

We went down to London where I went in to see A/M Breadner again. Then Bruce and I conferred once more with Sir Frederick Bowhill and G/C Geoff McDougall. We gave them a detailed picture of the whole route, then took the train back to Prestwick. There were comfortable quarters in the Orangefield Inn at Prestwick sufficient for all our crews, coming and going; also, the maintenance plans with Scottish Aviation were working out well.

We were ready to leave on January 21, but the North Atlantic weather was bad. We flew Fortress 9204, with its new Tokyo tank fuel lines, down to Portreath in Cornwall intending to go back via the Azores. Then the weather on the Azores route turned sour. We waited at Portreath for a chance to get away until the morning of the 25th, when reports showed that the route via Iceland was opening up. Back we flew to Prestwick. Here I decided to return to Canada with F/L Bill Lavery and his crew, who had taken over 9204. The Atlantic head winds were too strong from the west for a direct crossing so we decided to return via Iceland. We were getting close to Iceland when the weather closed in again and we had to fly all the way back to Prestwick. We were 7 hours and 25 minutes in the air without accomplishing anything.

The next day, January 31, we tried again and this time we made it to the USAAF base at Keflavik in Iceland just before the weather closed in again. We were accommodated all together as a crew in a Quonset hut heated by a Coleman heater. Shortly after our landing a blizzard started to blow and our machine had to be weighted down by sandbags in addition to ground tie-downs. The storm lasted for days.

We finally got away on February 6, even though the aftermath of the storm left strong head winds on our route to Goose Bay. Everything was going well, in spite of low ground speeds, when after about seven or eight hours in the air Bill Lavery turned on the Tokyo tanks as the mains were getting low. Shortly after, he reported, "I hate to mention this, but the fuel doesn't seem to be flowing from the Tokyos." We were stunned, as we were back in the same situation we had encountered on our initial eastbound crossing. This time we were over extremely cold water where survival after ditching would be pretty well out of the question. We couldn't go into Greenland as it was socked in by weather. In addition, we were told that Goose Bay, which was supposed to have remained open was showing signs of closing in with snow. Our one course of action was to hold course for Goose Bay as long as we could.

We were watching our fuel gauges carefully. They would slowly go almost to the empty reading and then, for some reason, show a slight increase. This happened a number of times. Each time we would think that we were near to ditching when the gauges would move up a bit. Some fuel seemed to be getting into the main tanks, but not much.

This condition persisted, and although completely baffled, we were thankful that we were still in the air. Finally after 10 hours and 35 minutes we were near Goose Bay, with our fuel gauges still barely above zero. The aerodrome was now closed in with heavy blowing snow. Bill asked for some distress flares lighted at the approach end of the runway, in addition to the runway lights, even though it was still daylight. On the first approach he just missed the runway but we saw the flares. On the second, he was lined up nicely and did a beautiful job of putting the aircraft down. Were we ever pleased to be back on the ground!

We were stuck with weather the next day but managed to get off on February 8, direct to Ottawa. We had refuelled so had sufficient fuel without using the Tokyo tanks. We landed at Rockcliffe after 5 hours and 15 minutes.

24

Moving the Wounded

It was great to be home again. Lin was as pleased as I was about the OBE and promotion. She had purchased a group-captain's hat for me, with its gold peak. It fitted perfectly. The trip, though long and at times hazardous, had paid off. The mail service was well under way and having its effect on the morale of the troops.

The engineers at Rockcliffe had gone over 9204 to locate the second dose of fuel trouble. Apparently Scottish Aviation had fitted new fuel lines from the Tokyos to the main tanks to replace the lines that had been sawed off. They had not noticed, however, that the ends of the vent lines from the Tokyos, which allowed air into the tanks as the fuel flowed out, had also been crimped or plugged. This meant that fuel still could not flow. However, one of the rubber inspection caps of the Tokyos had a long hairline crack. Apparently, when our main tanks got very low, the hairline crack in the inspection cap would open and allow enough air in for some fuel to flow to the main tanks. The cracked cap was the only thing which let us get to Goose Bay and avoid ditching. I have since kept the cap in my den as a souvenir.

During my absence, Marlowe Kennedy had kept everything going smoothly. The Ferry Wing, Domestic Transport Wing and now Overseas Wing were all in full operation. I made plans to move Bruce Middleton from Commanding Officer of 168 (Heavy Transport) Squadron up to the more senior command of the Overseas Wing. But first we had to find a replacement for him. In due course we managed to get hold of W/C Len Fraser, a pre-war commercial pilot who was home after a tour as Commanding Officer of 405 (Bomber) Squadron.

The Chief of the Air Staff, A/M Leckie, pleased with our Overseas operation, was planning a trip to the West Coast and back. As I wanted to look at our transport operations there and in Edmonton, it was arranged that I fly him. We took Lodestar 563, with co-pilot F/O Westman and crew, and left Ottawa March 13, arriving that night in Winnipeg via Kapuskasing. On the 14th we arrived in Vancouver after a refuelling stop at Lethbridge. Most of the mountain trip was at night and on instruments in thick weather. During our Lethbridge stop my mother and sister came out to the airfield to say hello. My brother Eric, now a leading aircraftsman and stationed at the Lethbridge airfield, also came over. I was able to introduce them all to A/M Leckie.

On a side trip up to Edmonton I saw W/C Hump Madden who so successfully was moving a lot of personnel and equipment on his 165 Squadron North-West Staging route operations. On March 24 we got back as far as Winnipeg and on the night of the 26th were again in Ottawa. Len Fraser was now well acquainted with his new command. When the Normandy invasion took place in June 1944, we needed an air service to move Canadian personnel and supplies to our invading forces, while bringing wounded out to hospitals as fast as possible. We set up No. 168 (Heavy Transport) Squadron detachment in the United Kingdom, equipped with Dakotas which our own pilots flew over. It was established at Biggin Hill aerodrome under Ronny Knowles operating between the United Kingdom and the Continent.

On August 4, with Len Fraser as captain, we took off from Ottawa on an overseas flight, leaving Marlowe Kennedy as acting commander of the whole operation. We flew to Gander in Fortress 9205 arriving in 4 hours and 55 minutes. On the 6th we chose the Azores route, due to weather, arriving there after a night flight of 7 hours and 20 minutes. At one point along the way, we listened in to radio conversations between commanders of German submarines. Apparently they were surfaced somewhere in the blackness below and were having a chat. Of course, we couldn't see them or do anything about them. 9205 had a long-range fuel tank in the bomb bay which developed a small crack. Because of our high altitude, the fuel started spraying all over, including into electrical equipment in the bomb bay. We shut off most electric circuits and descended rapidly to reduce pressure in the tank. We all chewed gum as fast as we could and spread it on the leak, having to hold it there until we reached a lower altitude. Meanwhile, the gasoline and fumes lingered for some time in the aircraft—we were quite literally a flying bomb until it eventually dissipitated. It had been a frightening experience. We refuelled and had something to eat in the Azores, then headed off again for Gibraltar, where we landed 6 hours and 35 minutes later.

We were stalled by United Kingdom weather until the 9th, when we took off for Prestwick, arriving in very bad weather. Len Fraser twice tried his landing at Prestwick, but there was hardly any ceiling. We gave up and flew to the RCAF's bomber station at Linton-On-Ouse, where we managed to get down. We were worn out after 10 hours and 10 minutes aloft. The Station Commander turned out to be my old friend G/C Clare Annis, who gave us the best of care.

We left our Fortress at Linton-On-Ouse after arranging for a first pilot to come down from Prestwick and pick it up. Len and I jumped on a train at York and went to London to see A/M Breadner. We found ourselves a double room on the top floor of the Savoy. We realized later that wasn't a good place to be as London was then suffering the onslaught of Hitler's V-1 flying bombs the "Doodlebugs." The V-1s came over at any time, night or day, so that sleep was difficult.

It was good to see A/M Breadner again. We sat in a room behind his office and enjoyed a drink while he described the situation. The V-1s were at their peak and causing severe damage and casualties. A large-scale evacuation might be necessary unless things improved. While we were talking the "Immediate Danger" bells rang through the building, warning of a Doodlebug heading our way. I wanted to duck under the heavy table between us, but the Air Marshal didn't make a move so I couldn't either. The Doodlebug passed over low and exploded somewhere beyond.

The Air Marshal told me that the matter of setting up our proposed air transport service onto the Continent from the United Kingdom was complicated, as the Air Commander in Normandy was against any additional people going into the war zone for the time being. He said he would try to set up a Normandy trip with the permission of the Marshal of the RAF, Lord Portal, as soon as he could. Len Fraser alerted Ronny Knowles, to have Dakota No. 971, ready to leave Northolt aerodrome as soon as clearance came through.

While awaiting further developments, Len and I kept our double room in the Savoy. The V-1s continued to thunder over, seeming to barely clear the roof before exploding beyond in the city. During one of these raids, W/C S.R. "Stan" McMillan, lately of Mackenzie Air Service in the Northwest Territories, telephoned. He was on his way to Ceylon to take command of No. 413 Squadron. We drank and talked through the afternoon and most of the night, getting pretty well inebriated in the process, to the point where the Doodlebugs and the occasional V-2 rocket didn't seem to matter. I recall stumbling into the bathroom and Stan holding my head while I was sick. We saw Stan on his way in the morning and Len and I staggered back to the Savoy.

A/M Breadner had arranged a reception at his quarters in north London, at which Lord Portal and other senior people were to be present. I was picked up by G/C Quiller Graham, the Air Marshal's chief assistant, and taken out to the quarters. Len Fraser and his crew, meanwhile, were standing by at Northolt. During the evening, Lord Portal telephoned the Air Commander in Normandy about other things, but then urged that our Normandy operation be approved. It was. I left the gathering immediately after assuring Breadner that we would be airborne from Northolt by 8 a.m. The idea was to get in there before the clearance was retracted.

We took off early on August 15, flying below 3000 feet; otherwise we were in danger of being shot down. On the way we flew over a great fleet of warships and transports, many of which were flying barrage balloons for protection. We made our landfall on the French coast and flew on over a smashed, dusty countryside, with solid rows of trucks moving to the battle areas, loaded with supplies and coming back empty.

We found our airfield, number B.14. It was a bulldozed landing strip used by a Typhoon squadron. We landed in a shower of dust and pulled off to one side. There were tents where squadron personnel lived all around the field. Typhoons were taking off and landing continually and were quickly re-armed with rockets after each landing. The gap had been closed at Falaise and a great deal of German armour was trapped there. The Typhoons were smashing up the German tanks with their rockets as fast as they could. From the strip the steady rumble of the guns up front could be heard.

We were met by a number of medical officers, including a brigadier, who were very interested in our proposed operation. They took us over to a field of wounded personnel lying on stretchers in rows and covered by the dust that smothered everything in the area. They couldn't evacuate them fast enough and many of the wounded urgently needed proper hospital care. I had been under the impression that the wounded, at this stage, would be in national groups of some sort, but when I asked where the Canadians were, the brigadier shrugged and said they were scattered throughout the area. My instructions had been to ensure that Canadian wounded were taken care of, but when I saw the situation it was obvious that the best thing was to forget about nationalities. The brigadier agreed to fill our aircraft with the worst cases regardless and also ensured that our crews would be looked after. While there, I bumped into F/O John Bonner, an RCAF administrative officer whom I had met previously and who was stationed at the strip. He agreed to keep an eye on things from an RCAF point of view.

The first load of badly wounded was brought aboard and their stretchers locked into place in tiers in the aircraft. We were to take all our loads to a place called Blakehill Farm in the south of England where casualties were sorted and transferred to other hospitals. We took off in the afternoon and swung around in a wide sweep to have a look at things, then set course for England. Back we went over The Channel, the ships and the balloons.

Our wounded were all in pretty bad shape but those that were conscious were as cheery as possible. One case was a British soldier, a diminutive Cockney, who had been badly wounded in the chest. The medical officers felt he wouldn't survive but were giving him a chance anyway. I chatted with him and assured him he would be at Blakehill soon. He grinned and said, "Sir, I'll be all right, I'm going home." He was still alive when we landed and I imagine, with his spirit, he probably recovered.

Two aerodromes near each other showed up in the thick haze. By accident we landed at the wrong one, so had to take off again. Once down on the proper field, our load was taken away quickly and efficiently in ambulances. We confirmed the arrangements that this base should be the British unloading point for our service and then took off and flew back to Northolt.

August 16, was spent at A/M Breadner's headquarters, going over with him and G/C Gordon Truscott the arrangements that had been made in Normandy and at Blakehill. They guaranteed from then on that the support so necessary for our operation would be given. The ambulance and troop supply operation followed the movement of our troops right across Europe, moving great numbers of wounded and supplies successfully. S/L Ronny Knowles directed this job with great competence.

Another visit to Blakehill Farm was required to finalize our plans. At the same time, I wanted to go to Leicester East to visit a group using special air supply dropping equipment, panniers, parachutes, etc., as I felt that something of this sort might be useful later on. I bid goodbye to the Air Marshal and G/C Truscott. This was the last time I was to see Breadner as he died later back in Canada. He had always been a great help to me since the early thirties.

On the 17th with Len Fraser in the captain's seat, we left Northolt in Dakota 972 and flew to Blakehill Farm. Here we completed our business and took of for Leicester East, where I had a good look at the supply dropping equipment and also some free-dropping panniers. There, we learned that an RCAF Dakota being ferried over the Atlantic by W/C Ray Goodwin had had a cockpit fire. Ray and his crew had subdued the fire but were left with little in the way of electrical instruments and equipment. They were being directed to land at Leicester East on an emergency basis. We were unable to wait to see Ray arrive so we left a message of congratulations on the great job he had done. Ray later joined Air Transport Command, then became Regional Director of the Department of Transport in Toronto.

We left Leicester East and landed in Prestwick, planning to return to Canada via Iceland, but after two days in Prestwick it was obvious the North Atlantic route was unstable weather-wise. We boarded a Dakota captained by F/L C.H. Ready flying back via the Azores.

We took off from Prestwick heavily loaded with mail, and headed for St. Mawgans in Cornwall, where we landed 2 hours and 20 minutes later.

That evening we watched a gallant young pilot land his badly shot up Mosquito. He had lost a large piece of one wing, so the controllers had diverted him to St. Mawgans which had a long runway. He made a first-class emergency landing, saving both the aircraft and crew.

At dawn the next morning we were awakened with a message that one of our Fortresses en route from Gibraltar to Prestwick had collided with a Wellington over the Bay of Biscay while flying in cloud. The Wellington had crashed and the badly damaged Fortress was staggering towards an air strip, Predannack, in southern Cornwall. The crew had thrown their mail load overboard along with personal kit bags and anything else they could in order to lighten the aircraft but it was still having difficulty staying airborne.

Len and I borrowed an RAF car and driver and set off for the landing strip. When we arrived the Fortress had landed and the crew was just disembarking. The aircraft was the worst-looking wreck I have ever seen yet was still able to fly. Only one propeller was undamaged, two had bent blades and the fourth had its blades sheared off at the engine cowling. The bomb aimer's compartment was crushed open, pitot heads wiped off, and pieces of the Wellington were embedded here and there in the leading edges of the Fortress. The aircraft had apparently struck each other a glancing blow. We put F/O A.B. Hillcoat and his crew on a train to London for a week's well-earned leave and made arrangements for repair of the aircraft. All of the crew were later decorated.

We left on the 22nd for Gibraltar. The following morning F/L Proudfoot, in Dakota 653, was ready to leave with a load via the Azores, so Len and I jumped aboard. After 8 hours and 30 minutes, we rolled along the metal strip runway at Lagens in the Azores. An hour later F/L McElrea was ready to go to Gander in Fortress 9203, so Len and I joined him for a 9 hour and 40 minute flight. The next morning, August 24, we flew to Rockcliffe in 6 hours and 10 minutes. There I found Marlowe Kennedy had everything running smoothly in the Command, and my superiors at Headquarters agreed with my decision to carry all nationalities of seriously wounded in Normandy. Changes were also in the air. Earlier in the year, plans had been made to replace our Fortresses on a gradual basis with new Liberators. These aircraft were now becoming available and our first deliveries were due in a month or so.

25

Going Regular

The Quebec Conference of September, 1944 was attended by most of the Allied leaders including Roosevelt and Churchill. It was held at the Chateau Frontenac in Quebec City. I was instructed to provide VIP air transport for personnel and important despatches.

Two rooms were provided in the Chateau, one for staff living accommodation and the other for an RCAF Air Transport business office. I sent our top air traffic specialist, S/L Leslie Collins, to run the Quebec operation. F/L Dick Forrest, at our Ottawa headquarters, was on the other end of the circuit. Demands for special flights came thick and fast as we supplied transport for almost all nations taking part, except the United States, which looked after its own requirements. Crews, including flight stewards, were operating night and day for some time. A special pass authorized by the Commissioner of the RCMP was provided for me to enter the Chateau. I flew down on September 19 and returned on the 21st, having met many of the senior military people and had a good look at that end of our operation. Many flights were carried out by our crews during the conference, everything coming off smoothly.

In October the first of our B-24 Liberators arrived. Len Fraser had himself checked out on the aircraft right away; when he was qualified I had him check me out, which took part of two days. I liked the Liberator although it handled a bit like a heavy truck on landing and take-off. It was faster and more reliable than the old Fortress, so naturally the pilots preferred it.

It was about this time that No. 168 Squadron engineers under S/L Bill Lewis modified a Liberator with a sleeping cabin in the rear. The aircraft was flown overseas and delivered to A/M Breadner as his personal aircraft. He did a lot of travelling to the various war zones and an aircraft was needed in which he could be comfortable.

My old friend A/C John Plant had returned from overseas and was now serving at Air Force Headquarters as Deputy Air Member, Air Staff. His chief, the Air Member, was A/V/M W.A. Curtis (later to become Air Marshal and Chief of the Air Staff).

John and I flew a number of trips together, mostly to the United States. One trip was to decide whether we should purchase a number of United States Navy RY-3 transports. The RY-3 was a single-tailed Liberator. We left Ottawa on November 24 in a twin-engined Beechcraft Expeditor with John as captain and I, co-pilot. We over-nighted in Washington, and in the morning flew to the big US naval base at Patuxent, Maryland. With us was A/V/M George V. Walsh, the Air Attaché in Washington.

We looked over the RY-3 very carefully and flew in it with a U.S. crew. During the demonstration flight we learned that under heavy icing conditions the tail would shake so much that the plane was almost unmanageable. This was enough information. We knew the RY-3 would not suit our requirements.

Later in the day, we were offered a chance to ride in a helicopter that was being tested by the US Navy. My old aversion to helicopters or autogyros, plus the fact that I was horrified at the thought of riding in something I couldn't fly, came to the fore. I adamantly refused. I could imagine nothing worse than being caught in the air in such a machine, then perhaps having the pilot pass out, leaving me sitting there unable to fly it! The two senior officers were annoyed at me but I would not go. A year or two later I was involved in a search and rescue operation which required me to fly in a helicopter. This helped cure my aversion to helicopters. We returned to Ottawa on December 2.

A little later my dizziness-and-collapse syndrome hit again. G/C Martin Costello and I were to appear at a late cocktail party in the Chateau Laurier, hosted by Prime Minister Mackenzie King. We worked late, there being no point in going home and then returning for cocktails. At the Chateau we entered the large room, filled with people, where the party was being held. I recall walking in with Costy and meeting one of the McDonough brothers who was involved with an aircraft plant in southern Ontario. A waiter offered us drinks, but before I had a chance to take a swallow, everything went black.

I came to in one of the Chateau's rooms in the care of G/C Ed Hall, one of our senior medical officers. (Ed Hall later became the president of the University of Western Ontario.) He had me rest until later that evening when I went home. I had been working too hard for some time. The next day saw me back to normal.

In January 1945 the Directorate of Air Transport Command was ordered reduced to the status of a group, with headquarters at Rockcliffe. By February 1 the major part of the job was done. We were now called No. 9 (Transport) Group with me as the first Officer Commanding. A suitable headquarters building was taken over at Rockcliffe and the staff of No. 9 (T) Group moved in. Our two ferry squadrons were disbanded; No. 170 Squadron in Winnipeg went first, followed a little later by No. 124 at Rockcliffe. Some of our smaller transport detachments across the country were also closed. We accustomed ourselves to the somewhat smaller operation and soon everything was ticking smoothly again.

In March A/V/M Curtis and I flew to Cartierville, Quebec, for discussions about a new transport aircraft for both civil and military operations which was to be built by Canadair Ltd. — the famous North Star. It was to use Rolls-Royce liquid-cooled engines. I objected, saying I would like to have Pratt and Whitney air cooled engines installed. Both Curtis and my old friend from Trans-Canada Air Lines days, J.T. Bain, said I was wrong and that the Rolls-Royce engines would be better. Time proved them right. The Rolls engines did a terrific job, particularly under severe Arctic conditions. Late in the war, plans were being made in Canada for a Pacific contingent called "Tiger Force." It would come into being when hostilities ended in Europe. John Plant and I were to discuss certain aspects of Tiger Force with our American counterparts in Washington, D.C. We flew an Expeditor to New York on April 16, where we were stopped by weather. On the 17th the forecast was none too good but we took off and headed for Washington. The weather had almost closed in when we arrived so we found ourselves placed high up in a stack of aircraft awaiting a chance to land. We flew on instruments in ice and heavy static, back and forth and were simply to keep waiting. But our fuel by now was quite low and the only place within our range which gave us a fighting chance of landing was Harrisburg, Pennsylvania. Everything else was completely closed in. The airport we had in mind was a strip with one end on the bank of a river and the other butting up to the base of a hill about 1300 feet above sea level. It was now night and we were solidly on instruments as well. The let-down, for which we had the necessary radio range card, was to be down the face of the hill at a fairly steep angle, then a quick flattening out and touchdown, followed by firm braking to ensure that we didn't go off the end of the runway into the river. I now took the left-hand seat. We flew the letdown accurately and then on the final leg started our steep descent, occasionally seeing the tree tops on the side of the hill in our landing lights. We braked to a stop short of the river and taxied to the ramp. A very tricky letdown!

We were welcomed by US Navy personnel from a photographic unit stationed there. They reported that under such weather conditions most arrivals had to be fished out of the river. In the morning we could see the hill and the runway in daylight. "Good God!" was all we could say. We took off again for Washington, having no trouble this time, completed our business and returned to Ottawa. Then V-E Day arrived. Victory at last. It was difficult to realize that the war in Europe was over.

A/C Plant officially took over as Air Officer Commanding No. 9 (Transport) Group about this time, and I became his Chief Staff Officer. It was a good arrangement as John was a fine air transport officer and pilot with whom I could easily work. He brought with him his personal staff officer, F/L Butch Handley, also a good type for our operation.

One of the first things John did on taking command was to get himself checked out as captain on the Fortress. At this time, a large quantity of penicillin was required in Poland. The government of Canada supplied it, with No. 168 Squadron to provide delivery. The first shipment was flown over successfully, but on the second flight Fortress 9202 crashed in Germany on November 4, 1945. Its five-man crew died. John Plant successfully delivered the third load of penicillin to Poland.

The first meeting of what was to be the United Nations was arranged for San Francisco. Diplomats of many countries were to attend. The Group was instructed to set up a VIP military service between Ottawa and San Francisco for the duration of the meetings. I was to make a survey flight of the route. With F/O Chaster as co-pilot and with our VIP traffic expert, S/L Leslie Collins, I took off in Expeditor 1389 on May 2, 1945. We flew to Windsor on the first day to make arrangements for our service requirements there. The following day was lost due to bad weather. On the 4th we flew to Chicago, where I met with O.T. Larson, one of our executives in the early days of TCA. He was now a vice-president with United Airlines.

I told him of our proposed service, of its fuel, maintenance, hot meals aloft, despatching, weather and other requirements. He immediately made arrangements at all our planned stops through to Oakland, California. Our stops were made at United Airlines bases where first-class service was to be the rule throughout the operation.

The following day we flew to Omaha, refuelled, then went on to Cheyenne, Wyoming, where the weather blocked us again for a day. On the 7th we landed in Ogden, Utah, and then flew on to Oakland, the terminus of the run. We stayed in the Oakland-San Francisco area long enough to see our planes arriving and departing on the service. I had some work to do with our units in western Canada, so planned the return trip that way.

We left Oakland on the 11th, landing en route at Medford, Oregon; Seattle; and Paine Field, Washington; reaching Vancouver the same evening. On the 15th, after completing my work, we took off for Prince George, Edmonton and Lethbridge, where we spent the night. On the 16th we arrived in Winnipeg after a stop at Rivers, Manitoba. On the 19th we were back in Ottawa. Our air service to San Francisco was now in full operation and was to conclude with usual efficiency.

Early in August I was sent to Bryan Field, Texas, for training in a new system of instrument letdowns called Ground Controlled Approach. I was to make a full report, with recommendations for or against the system.

With W/C Maurice Lipton as co-pilot, I left in Expeditor 1389 on August 7 and, after refuelling at Detroit, night-stopped at St. Louis, after six hours of flying. On the 8th we refuelled at Texarkana, Texas, then arrived at Bryan Field after 5 hours and 10 minutes in the air. On stepping from the aircraft at Bryan Field I was met by the Station Commander, Col. Duckworth. It turned out he had been an airline pilot before the war and we had met while both civilians. He explained that he had a very special course planned for me and asked that I abide fully by my instructions. He wanted to prove to me, beyond any shadow of a doubt, which instrument approach system was the best available.

Next morning found me on the flight line bright and early. I was to fly a Flying Fortress, using the SCS 51 (known today as ILS) blind approach system. The rest of the crew was thoroughly trained in its use. We flew all morning shooting blind approaches on SCS 51. In the afternoon I was given ground school in which every detail was thoroughly covered. By the end of the day, I was worn out but was told to be on the flight line at eight o'clock the following morning.

When I reported in, there was my Expeditor warming up, and sitting in the right seat was my Fortress instructor of yesterday, Capt. Potter. He explained that we were going to use the new Ground Controlled Approach system and that I would be given no instruction of any kind but merely do as I was told. A special coloured material had been fitted around the inside of the cockpit windows and I was provided with a pair of coloured goggles. Without using the goggles, I could see through the windscreen, but with them I could see only the cockpit instruments. I was to do all the flying while Potter rode as safety pilot.

I was to take off with my goggles up. I did so and on climbing out a voice on my radio, using the identification "Mica," told me to pull my goggles down. I was now flying on instruments. Capt. Potter said nothing, but the radio voice gave me compass heading and altitude. I turned onto heading and climbed to the altitude given. I knew I was heading away from Bryan Field. I was then instructed to turn right for identification purposes. On completion, the voice said I was now to carry out my orders accurately. I was given a new heading and told to lose altitude at a given rate. We were obviously doing an instrument approach into a field I had never seen before. Down and down I went with small corrections to heading and rate of descent. I knew I was getting close to the ground and starting to worry when I was told to flip my goggles up. I was almost down on a long runway with a white line right down the middle. I only had time to ease my control wheel back and complete the landing, straight ahead. I was impressed! Without any instruction, other than the ability to fly on instruments, I had let down on instruments almost to ground level on a field I didn't know existed, a new field under construction. This was the kind of instrument approach system we were looking for.

We taxied to the Ground Controlled Approach vehicle parked beside the runway and were given a general idea of how it worked. After that we shot a few more approaches, then returned to Bryan Field, where Col. Duckworth was waiting. I expressed my enthusiasm for the new system, and next morning I was put back in the Flying Fortress for three more hours of flying on SCS 51. After the ease of yesterday's practice, my three additional hours of drudgery on SCS 51 convinced me beyond a doubt that GCA was the only system for the RCAF. Duckworth only laughed at my enthusiasm, saying he had intended to make sure that I was well convinced before I left.

On the morning of August 12 W/C Lipton and I headed back to Ottawa via Wichita and Chicago. We arrived home the night of the 13th. I submitted my report, strongly recommending adoption of the Ground Controlled Approach system. My recommendation was accepted and GCA later became standard RCAF equipment.

On August 14 Japan surrendered, ending the Pacific War, and also ending our Tiger Force plans. The world,. for a short time at least, was at peace.

At the beginning of October I was asked to assist with the Provisional International Civil Aviation Organization, the aviation branch of the United Nations. Its headquarters were being set up in Montreal under Dr. Edward Warner, the president. A Canadian delegation was being formed, headed by Anson McKim of Montreal, and I was to be McKim's chief advisor. A room was arranged at the Windsor Hotel, and from early October until December I flew back and forth to Montreal, where I spent about half of each week. On my return to Rockcliffe I would catch up on my Air Force work. It was nice to see A/C/M Sir Frederick Bowhill again, as he headed the British delegation, and also G/C Tiny White, the head of the New Zealand delegation. By December we had a functioning Canadian delegation so I returned to my Air Force duties full time.

At this time it was suggested that I return to civil aviation where a good positions awaited me through the kind offices of C.D. Howe. The RCAF indicated that if I wished to stay on in the Regular Force, it was likely I would be accepted. Lin and I discussed the matter carefully, concluding that most of our friends were now Air Force people. During our five-years away from civil aviation, many of our friends there had moved away or retired. We decided not to make a major change in our lives again and to stay with the RCAF. We never regretted the decision.

26

Arctic Ops and a VIP

Group operations continued busy but uneventful though some changes in the senior officers had occurred. W/Cs Marlowe Kennedy, Bruce Middleton and Len Fraser had returned to civil aviation. W/C Donald E. Galloway (formerly of TCA, then 11 (BR) Squadron) was now in Group headquarters and W/C Ray Goodwin had replaced Len Fraser in command of 168 Squadron.

Between Christmas and New Year 1945, I had to make a quick flight to Winnipeg in a Lodestar with F/L Benny Gruenwald as co-pilot. We were planning to get back to Rockcliffe for the New Year's Dance in the Officers Mess, an event we had promised our wives. On the return journey, we ran into bad weather at Kapuskasing, where we stopped to refuel. I phoned Lin from Kapuskasing, telling her to go on to the dance where I would meet her. We left Kapuskasing in IFR conditions, and after battling ice and static, made it to Rockcliffe. Everything was normal until we touched down. The wind was across the icy runway and as our aircraft slowed, it slid sideways across the runway, coming to rest in a snow bank, though no damage was done. Benny and I hurried up to the Mess and joined the ladies just before the midnight celebrations bringing in the New Year.

Besides regular group operations in early 1946, we began planning for "Operation Musk-Ox," in which we were to drop fuel, food and other supplies from the air to a moving ground force in the far north. This force comprised a number of tracked vehicles, manned by specially trained Army personnel. They were going to drive from Churchill, Manitoba, across the snow-covered Barren Lands to the Arctic coast, across the ice of Queen Maud Gulf to Coppermine, on to Great Bear Lake and Fort Norman, then south to Fort St. John, BC, and thence to Edmonton.

We trained air drop teams, using special parachutes and containers.

Under certain circumstances we would use transport gliders which could land with a load and be snatched back into the air by the mother aircraft. The US loaned us two wartime Waco gliders, a C-47 equipped with the snatch gear, and crew. The first demonstration was to be at Rockcliffe on January 19, 1946. I was to fly in the glider, which was not a happy thought but it had to be done. The Waco was sitting in the snow and ahead of it were two small uprights between which was stretched a nylon cord arrangement which extended back to the glider as a towing cable.

The glider pilot was S/L Somerville of the RCAF. His co-pilot was A/V/M Art James. As the Waco had no seats, the rest of us were *standing* in its cabin, holding on to the fuselage tubing — a bit crude but effective. We soon heard the C-47 diving over us. It had a large hook and cable running up to a powerful winch in its cabin. The C-47 snatched our cable, there was a terrific jerk and we were airborne with practically no run at all. We hung on grimly and a bit later, when sufficient altitude had been achieved, S/L Somerville pulled a lever to drop the cable and we were on our own. He glided the machine around very nicely, then landed routinely. It had been a frightening, yet interesting experience.

In a weak moment I agreed to fly the glider myself during a snatch-off. Somerville would let me know when we would fly, and I talked Don Galloway into coming along, as my co-pilot.

On January 20 Somerville called to say they were ready. I picked up Don and we drove to the site. There was the big glider with its snatch cables all ready. Somerville got us strapped in, gave me a few tips on gliding performance, then left us on our own. Don and I sat rigid in our seats waiting.

In a moment, the C-47 shot overhead. Its hook didn't quite catch our cords firmly. The hook had let go and our cords flew right back at us with a crash. I saw something coming and we ducked. The cords smashed the windshield in front of my face. We hadn't moved a foot!

Somerville inspected the windshield. I said, hopefully, that I guessed the flying was finished until the windshield was replaced. But no such luck — the glider would fly just fine with the damaged windshield. We strapped in again and our cables were put back in place.

Over came the C-47. This time we were firmly hooked and yanked from the ground in a flurry of snow. We climbed to altitude and cut loose. From there on I actually found it a most enjoyable experience, the glider being smooth and easy to fly and, of course, quite silent. There was no trouble in landing, as only a short landing run was needed. I was now completely sold on glider snatching, and so, I believe, was Don Galloway. S/L Somerville later did a glider flight or two during Musk-Ox, carrying drums of fuel, but we found in the end that our supply dropping technique was the more suitable.

Lin and I were house hunting again, as our lease had expired on Glen Avenue and the owner wished to move back in. Since we were staying in the Service, a house in Ottawa would be an asset, so we bought a new two-storey one in the Glebe. Lin was happy to have a home of her own again, although I am afraid it was not to last very long.

A/C John Plant, our Air Officer Commanding, was promoted to the rank of air vice marshal and posted to the West Coast to take command of Western Air Command. We were all sorry to see him go as he was a first-class airman and commander. His place was taken by A/C Larry Wray, effective February 15, 1946. I had known Larry for a long time, and knew I would get along well with him.

On February 28, 1946, our biggest squadron, No. 168 Squadron, was disbanded. It had done a fine job on the mail run to the troops overseas; its detachments which had done the ambulance and supply flights in Europe were also disbanded. The squadron had completed 688 flights overseas with the loss of three Fortress aircraft and crews. F/L Cathcart and his crew crashed at Prestwick and F/L Pat Hillcoat and his crew disappeared on a night flight from Rabat to the Azores. Aboard the latter flight was one of my very good senior staff officers, W/C Dave Wood. F/L Harling and crew crashed in Germany en route to Poland. An outstanding job of transatlantic flying during this period was done by F/L (later S/L) Wess McIntosh.

Operation Musk-Ox got under way from Churchill on February 15, 1946 and headed across the snow-covered Barren Lands for Baker Lake. I took a Dakota to Churchill, carrying A/C Wray, to have a look at our supply dropping operation. My co-pilot was F/L Brunet. We left Ottawa on February 25, refuelled at Kapukasing and night-stopped in Winnipeg. On the 26th we flew to The Pas then on to Churchill in very cold weather.

We were given quarters in one of the long, wartime huts. At this time of the year, they were almost completely buried in hard, drifted snow, so it was only possible to get in and out through a doorway at each end. In our hut also was the Army Base Commander, Col. Donald Cleghorn, and the senior Army representative, Col. J. Tuzo Wilson, otherwise known as Jock. Jock's father was J.A. Wilson, the much respected Director of Civil Aviation. We had some worthwhile discussions about the north with Jock and Don, both of whom had plenty of experience with it.

We had our meals in a combined mess hall. At meal time, we lined up, each with his US Army type of compartmented food tray. The cooks placed the food in various compartments as we moved along. At the noon meal, we were eating from our trays when, as I swallowed a mouthful of baked beans, something crunched and I pulled out a small piece of glass. I had swallowed some other pieces. Across from me was a young RAF medical officer, a wing commander, who was a British observer for Musk-Ox. He took the sample of glass to the kitchen, returning in a minute to say that the wrist watch crystal of one of the cooks was missing, and I had had some of it for lunch! The MO had me eat lots of bread, and watched me carefully for 24 hours before declaring the danger period over.

The moving force of tracked vehicles was on schedule across the Barrens. The air drops had worked well initially, but during the last few days had been unable to make drops due to high winds and blowing snow which, in the Barrens, quickly reduce visibility to nil. The force had now arrived at Baker Lake but was out of fuel, and could move no further.

The weatherman suggested that a night operation to Baker Lake was the only hope for a few days — they expected the blowing snow to ease off that night. Since I had been into Baker Lake during my bush days, I decided to take the night flight myself. An ice strip was laid out there and would be outlined by gooseneck oil flares.

On March 4 I took off in Dakota 969 with F/L "Black Jack" Hall as co-pilot and a cargo of fuel drums. Visibility was poor but we managed to home on a temporary marker beacon at Baker. We arrived overhead in improved visibility. However, a crosswind was strong enough to extinguish many of the gooseneck flares. It was a difficult landing, but we managed it and kept from ground-looping on the ice.

While the troops were unloading the aircraft, we met with Col. Rowley, the Mobile Force Commander, and his officers, who briefed us on problems they had encountered. Back we went to our aircraft. We were rigid with cold. Our take-off worked out although we drifted slightly on the ice due to the cross wind. The landing back at base was O.K. until near the end of the run, when one brake froze, causing us to go off the runway, but with no damage.

The weather picture brightened and air drops were again carried out. A/C Wray had completed his tour so we left for Winnipeg, Edmonton and Yellowknife on the way back to Ottawa. We arrived home on March 13.

At this time I was awarded the Efficiency Decoration, a long-service decoration requiring at least 20 years service, and having been in the Army in September 1931. It so happened that my commission was not transferred from the Army militia to the Air Force until late 1931, so I qualified for the E.D. It was presented on parade at Rockcliffe by the Commanding Officer of the Station, G/C R.F. "Pat" Gibb.

Field Marshal Viscount Montgomery of Alamein was coming to Canada in August and would tour Canada, plus visit the US. I was to be the air commander of the tour which would use two Dakotas; and Col. Harold M. Cathcart of the Army was to be in charge of administration. Lieut.-Col. Trumble Warren of Hamilton, formerly Monty's aide for part of the war, was returned to duty as chief aide. Lieut.-Col. "Pot" Doucet (later Brigadier) was in charge of PR and press, under Harold Cathcart. F/Ls Ed Wilson and C.S. "Stu" Olsen were to be captains of the Dakotas.

Many planning sessions were held to determine tour stopovers, social arrangements, inspections of troops and veterans and so on. Arrivals had to be accurate as Monty was going to be met at each stop by important officials. I actually had authority to break air regulations as necessary to keep to schedule.

Our two Dakotas left Ottawa on August 23, 1946, and flew to Halifax in 3 hours and 35 minutes. Monty, his party and ours were to put up at the Nova Scotian Hotel.

Monty's ship arrived the next day and we were all at the dock to meet him. After the inspection and greetings, we returned to the hotel, where we went over Monty's travel plans with him. He stressed the importance of his seeing the war veterans at each point. Wherever there was a military hospital on his route, he wished to visit it to see the disabled veterans. We later accompanied him through these hospitals, where he spoke to each man, especially those who were bedridden. In some cases these courageous men had been lying in hospital since the First World War. Once, Harold Cathcart and I were invited to Monty's room. He had changed into comfortable civilian clothing and was spreading a series of maps on the floor. He showed us how his plans and operations had worked from Alamein through to the collapse of Berlin. It was one of the most interesting evenings of our lives. Monty was a first-class soldier, 24 hours a day.

The tour was thorough and must have been an enjoyable experience for Monty. It began for him with a round of ceremonies, including presentation of an honorary Doctorate of Laws at Dalhousie University on March 25, before heading west by air. At every stop, there were inspections of guards of honour, visits to veterans, formal luncheons and banquets with Lieutenant-Governors, premiers, mayors, etc. In Ottawa he attended a state dinner with Acting Prime Minister Louis St. Laurent.

The Field Marshal went out to Rockcliffe to inspect Air Transport Command Headquarters. While there, he had a long chat with Lin and Harold Cathcart's wife, Fran, who were quite taken with him, and I still prize pictures of the five of us together.

Not all the tour was by air; from Montreal to Ottawa, Monty travelled by train, and the aircraft flew ahead to meet him for the next leg. On September 4 we flew from Calgary to Victoria, giving him an impressive look at the Rockies en route. During the West Coast portion of his tour he met his brother, Donald Montgomery, a Vancouver lawyer.

Our aircrews and servicing parties were at peak efficiency, and everything went as planned. Having transported the Field Marshal and his party west, we carried them eastward again to Ottawa on September 9. After refuelling, we left for Quebec City. This particular flight was arranged so that Monty could spend the evening with Field Marshal Lord Alexander, the Governor-General, who was in residence at Quebec City.

The next morning, September 10, we were due to fly to West Point, New York, where Monty was to be received by the Commandant, Gen. Maxwell Taylor. The weather reports were bad, with low clouds, poor visibility, winds and turbulence, and in addition there was no suitable letdown system at our destination for these conditions. We decided we could make it by a combination of airline and bush type flying. Just before we left Quebec City, our crews were formed up into two small flights where Monty spoke to each man individually and thanked him for his services.

We departed on time, but from then on we had a real battle with the weather. The strong winds guaranteed that we would arrive late. We arrived at Stewart Field, West Point, two and a half minutes late, our one and only delay on the tour. The weather was so bad that West Point was certain we couldn't make it, but we flew low, just above the Hudson River, and at the right moment swung up over the bank and there was the field. It was apparent that we had caused some consternation by our arrival. We could see cadets running at the double to fall in for inspection, so we taxied in slowly. When everyone was in position, we taxied up and stopped. When Monty stepped out, he and Gen. Taylor had a real chuckle at the furor we had caused.

After the guard of honour inspection we drove to West Point and were entertained by Gen. Taylor at his home, followed by lunch in the West Point dining room with the cadets. In the afternoon we went to the chapel, where we listened to the huge pipe organ played by its organist and builder, Mr. Mayer. It was beautiful music, particularly when Mr. Mayer played "Lili Marlene" for Monty, a tune which had also been played for him on Ottawa's carillon.

Late in the afternoon, we said goodbye. We hated to leave Monty as we had all become very attached to him. We flew back to Ottawa, where we arrived, all very tired. I was proud indeed of the excellent job of flying and maintenance carried out by our RCAF crews. Col. Harold Cathcart and his Army team, who had looked after ground transportation, financing, accommodation and public relations, had also done a top job. All in all, the RCAF and the Army had worked together as a team to get the job done. Looking back, we had been a bit concerned about getting along with Monty, but in the end, we all enjoyed every minute of his company. With the tour at an end, we really felt that we were losing a very good friend.

27

Goose Commander

On returning to my position as Chief Staff Officer of No. 9 (Transport) Group, I was called in to see the AOC, A/C Wray. His news was that I was to be posted as Commanding Officer of RCAF Station Goose Bay, Labrador, one of our biggest and most isolated bases. I was to take over as soon as possible. I reported to the Chief of the Air Staff, A/M Leckie, who congratulated me on the success of the Montgomery tour, and told me that part of my job was to improve discipline and the general outlook at Goose Bay.

We sold the house we had recently bought to an official of the National Research Council, and managed to come out of it without losing money on the deal. Lin took on the job of selling most of our furniture. I packed my bag and once again said goodbye to her. As Goose Bay came under the Air Officer Commanding, Maritime Group, A/C Frank Wait, I had to report to him for further orders, so I flew to Halifax. On September 26 I left there for Goose Bay.

Henceforth I was busy familiarizing myself with the huge station, with its USAAF and DOT operations, as well as large RCAF establishment. As overall commander I moved through the various stages of tightening discipline and initially I was somewhat unpopular. Eventually, though, everyone appreciated being part of a hard-working, orderly station and morale improved accordingly. Up to now there had been no wives at Goose, but under new peacetime policies arrangements were made to have them arrive gradually. Old barracks buildings were modified into small apartments, and as each was readied a wife and family came to fill them.

At that time there was no Commanding Officer's house on the station, but a small building which had been the timekeeper's office and living accommodation during the construction period. It was small, but I had it cleaned up and some old furniture brought in. By December I was ready to send for Lin. My old friend Maurice McGregor of TCA brought her on one of his flights through Goose Bay (our wives were not permitted to fly on RCAF aircraft at that time), and she arrived just before Christmas.

I managed to get some flying done in our only airplane, a Noorduyn Norseman on skis. I enjoyed the machine as it let me practice my old bush flying techniques again. Meanwhile, I had become good friends with the USAAF senior officer, Col. Orie Schurter. The Americans could not have their wives at Goose Bay, so we often invited Orie to our home for dinner.

On January 12, 1947, an urgent request arrived for medical supplies to be flown to a remote bush post called Three Rapids. I intended to use our Norseman, but Col. Schurter asked that I take theirs, a Norseman on skis. But they had no experienced bush pilot. I agreed to fly their aircraft and check out their pilot, Lieut. Surcee. We flew to Three Rapids in an hour and I showed Surcee how to pick his landing spot and get in safely. We delivered the supplies and returned to Goose.

On January 29 we had another emergency trip with medical supplies and Dr. Tony Padden of the Grenfell Mission at Northwest River. We flew to a settlement up the coast called Nain. I had been checking F/O Barney Hartman as the station's search and rescue pilot. He was a sound, careful pilot, and was second pilot on the Nain flight. It was extremely cold, but everything went well that day and we arrived over Nain in two hours. The inhabitants had placed trees in the snow to mark a landing strip, so I showed Barney how to fly over and check the strip before landing. In this case the strip was a bit rough, so we flew around until I located a better place where we landed with no trouble. While Dr. Padden tended to his patients, Barney and I nosed about, meeting the Reverend Peacock who had long lived in the Nain mission, doing a wonderful job for the people.

Our return flight to Goose Bay was uneventful. Barney Hartman had got the idea of bush-type operations and from then on flew increasingly as captain, graduating to a Dakota when one arrived.

I was fortunate in having such a fine staff at Goose Bay. The senior administrative officer was S/L Robert "Sam" McBride, and the station adjutant F/L Newton Brydon. Our hardworking works and building officer was F/L Ray Baker, who doubled as fire chief. Flying control officers were F/L A.R. "Andy" MacKenzie (later shot down and captured in Korea), F/Ls John Buzza, "Buck" Bayley and others. James Wilson was our senior DOT official and Ralph Gillissie our local bank manager. S/L R.C. "Robbie" Bell was an outstanding senior medical officer who went on to great things as a plastic surgeon in England. Men like these made the whole operation tick.

In April 1947 I heard from the Minister of National Defence, Brooke Claxton, that I had been awarded the McKee Trans-Canada Trophy for 1946. This is the top civil aviation award in Canada. Needless to say I was surprised and pleased. Many of my old bush-pilot friends had been awarded the trophy earlier: C.H. "Punch" Dickins, W.R. "Wop" May, Walter E. Gilbert, Grant McConachie and others. I was honoured to join that elite group.

Messages of congratulations came from many old friends. The citation read, "In recognition of continuous outstanding performance as an officer and pilot of the RCAF, in both civilian and service aviation." I was pleased that the committee had thought of me that way, but I had only done my duty to the best of my ability, and had enjoyed doing it. No more, no less.

Barney Hartman was a skeet shooter and set up a skeet range at the edge of the station, using surplus material to assemble everything but the hand traps. Frequently Lin and I would accompany him and his wife, Jo, to the range in the evenings to help by operating the trap. He quickly became a crack shot and rose to be the world champion in all or almost all categories. Later he became an executive of Canadian Industries Ltd. in Montreal.

Early in June Lin and I flew out for the McKee Trophy presentation. The ceremony took place at Rockcliffe on June 11, 1947, before a large crowd, and scared me far more than the flights that had led to it. The Minister made the presentation, assisted by A/M Leckie. My reply was brief, and I was much relieved (and Lin greatly amused) when I was able to retreat to the Mess for one or two drinks.

Back at Goose Bay Barney had fitted floats to the Norseman, which was now moored to a buoy near the dock. No one on the station, besides myself, had ever flown on floats, and I was pretty rusty, not having been on floats since 1937. Barney and I got going on June 23 and 24, and after shooting a few circuits I checked him out. He was a natural and in no time at all was away solo.

The airfields at Fort Chimo (Ungava area) and Frobisher (Baffin Island) came within my jurisdiction. Frobisher was under command of an RCAF squadron leader, but there was a US detachment there as well, under an American officer not to exceed the rank of major. The USAAF had moved in a Lieutenant-Colonel, who thus outranked our man; there were also problems over the proximity of some Eskimos to the airfield. I decided to fly our Dakota to Chimo and Frobisher, taking Col. Paul Zartman, who had replaced Orie Schurter at Goose Bay, an RCMP officer, and one or two staff. We took off on July 9 and flew across the bushland to Fort Chimo in 3 hours and 45 minutes. After a good look at the base, which was deteriorating badly from disuse, we headed across Hudson Strait to Baffin Island and landed at Frobisher three hours later.

After an inspection there I asked Col. Zartman to replace the lieutenant-colonel with a major. As for the Eskimo camp, we decided that, for the good of its inhabitants, it should be moved further away. A better site was selected and a gradual move was later made which improved the situation. The next day, the 10th, we returned to Goose Bay, carrying two or three sick Eskimos for medical treatment.

During the fall we had a rash of fires on the base, something serious on a base 800 miles from civilization. The consequences could be disastrous should a fire get out of control. We suspected arson, and with help from special fire investigators sent in we finally apprehended the culprit. Soon things settled back to normal.

Lin and I flew out for a holiday north of Montreal. Following a good rest, I went to Trenton to meet with A/V/M E.E. Middleton, boss of Central Air Command, which now controlled Goose Bay. When we returned to the base, it seemed a comparatively quiet isolated corner after the bustle and noise of the outside world. In late November we hurried to Lethbridge, as my father had told me my mother was dying after a long, courageous battle with cancer. We stayed with the family for a week, and then had to return to duty. Mother passed away while we were en route to Goose Bay.

In December tragedy occurred when a USAF C-54, heavily loaded, took off from Goose Bay in a snow storm within the limits of ceiling and visibility. It crashed on a hillside a few miles away. Shortly after it had disappeared Col. Zartman called, so we alerted our base search and rescue organization pending further information. We had few details as the crash had occurred at night.

From a passing TCA aircraft came a report of a glow spotted through snow and cloud. The pilot gave his approximate position at the time of the sighting. Two ground rescue teams with dog sleighs were despatched with orders to be careful as there were streams with open water which could be dangerous in darkness. Aircraft were ready to go at first light if visibility improved.

Paul Zartman had recently received a helicopter for his side of the base and suggested we put it to use. With my strong dislike of helicopters I hesitated a bit but this was a real emergency.

At first light, despite continuing light snow, we were off, with Zartman and me as passengers and a young pilot at the controls. We groped through the snow until we could see the crash on a hillside. The ground rescue teams were just arriving after an arduous trek. The pilot selected a small frozen lake, encircled by trees but close to the crash site, and landed. From there Paul and I walked to the wreckage. It was a mess — 16 dead, nine injured. The survivors were patched up and taken out by helicopter and dog sled; later the bodies of the victims were brought out. Supplies were dropped from the air by the search and rescue aircraft. Much later Paul and I again inspected the site in the chopper; I still couldn't fly it but my dislike of the machine was fast fading. Henceforth I backed the helicopter, which had so graphically proven its usefulness.

In February 1948 I was notified of my posting from Goose Bay to Vancouver, where I was to be Officer Commanding, No. 12 Group. We crated our linens, dishes and clothes (our total possessions) and loaded them aboard the Dakota piloted by Barney Hartman. The day before departure my replacement, G/C Jimmy Verner, arrived at Goose. I showed him around; then, after a short ceremony and a drink together, the command of the base was his.

On the morning of March 17, 1948, a very cold day, we took our leave of the people of Goose Bay and departed. Every posting produced a few heart-tugs and this one was no different, for we had become attached to the people of the base, the Grenfell Mission and the Labrador region. After a flight of 5 hours and 20 minutes we landed at Rockcliffe. We travelled to Vancouver by train and just before the end of March, 1948, I took over the command of No. 12 Group. By coincidence, on April 1, 1948, my old transport organization, which had been reduced in size from the days of the Directorate of Air Transport Command to that of No. 9 (Transport) Group, was now made a Command, and thus the ultimate aim had been achieved.

28

12 Group Days

No. 12 Group was responsible for RCAF operations and administration in British Columbia, within the overall command of North-West Air Command in Edmonton, itself under A/V/M K.M. Guthrie. The group consisted of Jericho Air Station (where both 12 Group and Canadian Army headquarters and living accommodation were); the main flying base at RCAF Station Sea Island, commanded in turn by W/Cs Fenner Douglas and Marlowe Kennedy (back in the RCAF since July 1946): the RCAF side of Patricia Bay aerodrome (Vancouver Island, used mainly for air cadet camps plus training on flying boats and seaplanes); and some 15-20 Air Cadet Squadrons to which we extended supply and inspection services. The group flew Vampires, Dakotas, Beechcraft and Cansos, and also operated two rescue launches (HSLs).

Almost immediately we were involved in the annual air cadet inspections, each of which meant a flight to the squadron, inspection of the cadets and training quarters, and a lunch or dinner. This job was shared with my second-in-command, W/C J.D. "Red" Somerville, a distinguished wartime night-fighter pilot. During this period Red ferried our first Vampire from eastern Canada. En route he had to force-land in northern Ontario, causing serious damage to the aircraft and his back. Still suffering from his back injury, he flew a replacement aircraft to the Coast.

During the cadet inspections we normally took one of the group's aircraft and gave the cadets a "flip" if possible. Various members of the Provincial Air Cadet Committee would accompany us on these trips. They were dedicated men who devoted much of their time and money to the good of the young cadets; they included men like Irwin Finch, A.W. "Nick" Carter, a former First World War fighter pilot and Ontario Provincial Air Service pilot, and G/C Clare Tennant.

We were just completing the 1948 air cadet program when B.C. was hit with the most severe floods in its history. Large areas in the interior and almost the whole Fraser Valley were inundated. Rapid melting of snow in the mountains had triggered the floods, and damage was aggravated as dikes crumbled or overflowed. Houses, crops, livestock and people were all affected.

The Premier of B.C. declared a disaster and set up a Flood Committee with powers to take any action deemed necessary. The three heads of the local Services were on the committee — Col. Eric Snow of the Army, Cmdr. Bill Stacey of the Navy, and myself for the Air Force, plus one or two civilian members. From Flood Committee Headquarters we were all in communication with our Service headquarters, able to pass on orders for immediate action. The Army, backed by civilian agencies, supplied ground transport, the Navy provided boats and launches, and we supplied aircraft to move people out of certain areas and air drop sandbags on a round-the-clock basis. The sandbags were flown in from all over Canada and were dropped according to orders from our Flood Committee. Our Service headquarters was run by our seconds-in-command while the commanders served full time on the committee. Everyone worked nearly 24 hours a day to keep up the battle.

Between meetings I would fly members of the committee in an Expeditor over the Fraser Valley and Canyon. By getting a first-hand look at conditions we were better able to plan and coordinate operations. By the time the waters had begun to recede and the clean-up job had started, we were completely worn out. Finally, Operation Overflow drew to a close, and by July we were able to get back to more normal duties.

In September a US Navy Beechcraft disappeared between Churchill and The Pas. Aboard were Capt. Ben Custer, USN; Capt. Sir Robert Stirling-Hamilton, RN; Lieut. Wilcox, USN; Chief Petty Officer Kastner, USN; and Sgt. Scalise, US Army. The first two were the American and British naval attachés to Canada, and the subsequent search was dubbed Operation Attaché.

Soon after the plane vanished I went to Manitoba to command the search, turning 12 Group over to Red Somerville. On September 14th, I flew to Winnipeg where I was briefed by A/C Martin Costello, AOC, Tactical Group Headquarters, under whose overall command the search was taking place. The next morning I flew north to search headquarters at The Pas. There I was briefed by S/L Jack Hudson, who had started the operation. My second-in-command was Col. Sherwood Buckland, USAF, a friend from the American embassy. We had a large collection of some 35 aircraft and crews from all over Canada and the US.

We kept expanding our search pattern in all directions from the route the Beechcraft was to have taken. Aircraft flew every hour of daylight, changing crews if fresh crews were available, and flights went out at night looking for camp fires. I was pleased to regain the services of an old friend, S/L Jack Hone, now retired and in business in The Pas. He was taken on RCAF strength, given a Norseman on floats and a crew, and sent off into the bush to talk to Indians and trappers, gathering any information he could. When it was all over Jack was given immediate promotion to the rank of wing commander for his contributions, then retired again.

The operation lasted 12 days, with the search areas now expanded to what was considered the maximum fuel range, under any condition, of the missing plane. I was beginning to fear that it was at the bottom of a lake. Rather than cancel the search, however, I decided to expand further and continue for one more day.

On the 13th day a message came back from a Lancaster piloted by one of my West Coast pilots, F/O René Lemieux, that they had found the missing aircraft, crash-landed in a swamp, and then spotted the personnel a few miles away, heading through the bush. Lemieux circled them and headed them toward a nearby lake. A Canso, piloted by F/L Robert Virr landed there almost at dusk. I instructed them to remain overnight with the rescued people, as a dusk take-off could be dangerous owing to rocks in the lake. The Canso crew camped on shore with the survivors and gave them some light food.

They flew out the next morning with a Lancaster overhead to keep a motherly eye on them. We had doctors ready to look over the survivors on arrival. On hand was A/C Costello, who had flown in the previous evening and was thus able to see the finish of the operation. The Canso disgorged a party of bearded survivors who were thankful to be back in civilization. After medical examinations, some food and rest, they were flown out to Winnipeg. I broke up the search organization, sent all aircraft and crews back to their bases, and boarded a flight to Winnipeg, with Costello at the controls and Buck and I catching up on our sleep in the cabin.

The day after we landed, September 26, A/C Costello and I completed the disbandment of Operation Attaché, and on the 27th we flew to Edmonton for the annual Command Conference of North-West Air Command, presided over by A/V/M Guthrie, the AOC. That ended on October 1, after which it was back to Winnipeg, this time for a reunion of McKee Trophy Winners held in conjunction with the award of the trophy to Barney Rawson, an early TCA pilot.

That gathering was, I believe, the first of many annual reunions of McKee Trophy recipients, and what a gang of us there was to pose for a photograph! "Punch" Dickins, J.A. Wilson, Romeo Vachon, Pat Reid, Maurice Burbidge, Murton Seymour, "Wop" May, Tommy Siers, George Phillips, "Doc" Oaks, A.D. McLean, Barney Rawson and myself. As special guests we had Johnny Fauquier, a famous RCAF master bomber, and Jimmy Mollison, a noted transatlantic flyer. Also on hand were many old friends, including Frank Kelly, my engineer from the bush flying days. But, ah, what a hangover I had when it came time to fly back to Vancouver in my Dakota!

On my return I found a letter of congratulations regarding Operation Attaché from the Chief of the Air Staff, A/M W.A. Curtis, and thoughtfully attached were copies of messages from many notable Canadians, Americans and British.

After a short spell at home and at group headquarters, I was sent to a blind approach and landing test site near Eureka, California, to report on operations there. TCA was sending pilot Arthur Rankin as its representative, so we flew down together. The site was on a plateau in a mountainous area, with a camp and landing strip. All known blind approach and landing devices were being used, including FIDO, the wartime use of fire alongside the runways. Eureka had been chosen for its bad weather, particularly fog. Here the weather was described as "unflyable" when it was clear, and everyone rested or took ground classes. When the fog closed in, the sirens would scream and we would all hustle into flying gear and grope our way to the flight line, ready for operations. As the weather was unpredictable, we were always confined to camp, except by special dispensation. We made several flights on the various types of instrument approach and landing systems, but whenever the ceiling lifted to 100 or 200 feet the sirens would wail again and flying would cease. It was a world which operated in reverse, but it taught us plenty about the art of blind landings.

One evening the instructor in charge of the FIDO line invited Art Rankin and me to drive to Eureka for dinner. The meals in the camp were just average, so a good meal was welcome. Since he was a member of the staff and the weather was not closed in (it was only raining a bit), we were allowed to go. We soon wished we had stayed put.

"Mr. FIDO" had a powerful car and no sooner were we out of camp than he began to tear madly down the curving, hard-surfaced mountain roads, glistening with the rain. It was more terrifying than our blind landings. Finally I blurted out, "Mr. FIDO, we are passing signs which say 'slippery when wet'." Still driving fast, he replied, "I know, the signs have been there for a long time." We kept quiet thereafter, and had a fine dinner once we reached the restaurant, but had to face the same ordeal on the return journey. We accepted no more invitations!

After these experiments I went to Los Angeles to join G/C Max Hendricks and his group to examine the operational and technical aspects of a more advanced Ground Controlled Approach system, with Max and his party examining the technical angles while I tackled the operational ones. It was here that I first met David Callahan and his family. Dave was a senior executive of the Gilfillan Company, makers of the Ground Controlled Approach equipment. Our paths crossed several times thereafter.

From Los Angeles we flew to Ontario, California, where there was a long landing strip from which Gilfillan was carrying out its flying experiments. The equipment being tested was "automatic GCA" which could control automatically a number of aircraft on approach at the same time. On the morning of the 9th we returned to Vancouver and for the next two months I concentrated on No. 12 Group operations, doing much flying around British Columbia.

On February 24, 1949, I heard from A/M Curtis that I had been awarded the US Legion of Merit, in the degree of Officer. This was awarded by the President of the United States, on the recommendation of the US Navy, for our success in Operation Attaché. A/C Costello received the same decoration, while F/L Virr and F/O Lemieux received the Legion of Merit, degree of Legionnaire. I was pleased and honoured.

We four recipients converged on Ottawa on March 16 and were entertained royally, first at the American embassy, then at the residence of Ambassador Laurence Steinhardt (later killed in a plane crash), where the actual investiture took place. The investiture was impressive with many senior USAF and RCAF officers present, including the Chief of the Air Staff and A/V/M John Plant. Lester Pearson represented the government of Canada. We four guests of honour got through the ceremony well enough, though we had imbibed somewhat at the embassy beforehand, and it was something of a struggle to appear sober. On March 21 we departed by air for the West, feeling a bit ragged around the edges.

Back on the Coast, everything was going well, ably handled by Red Somerville, with the annual air cadet inspections going on. In May a small aircraft carrying a pilot and one lady passenger disappeared in a snow storm in the mountains. We quickly began a search and after considerable flying located the aircraft perched precariously near the top of Mount Hozomeen. The two people were spotted by Ray Munro, the flying newsman of B.C., apparently in need of help lower down the mountain.

Three of our parajumpers were dropped on the side of the mountain. It was a professional flying job by F/L Lyle Harling and a courageous undertaking by the parajumpers, led by Sgt. Leckie. As a result the stranded fliers were safely brought off the mountain slope.

About this time there were several changes in senior personnel. My Army counterpart in Vancouver, Brigadier Eric Snow, was transferred and his place taken by Brigadier Pat Bogert, who was just as cooperative as Eric had been. I also got a new boss, A/V/M Hugh Campbell, who became Air Officer Commanding North-West Air Command when Ken Guthrie retired to civilian life. Then my turn came. I was selected to attend Course No. 3 of the National Defence College in Kingston, running from October 1949 to June 1950. This course, highly prized in upper military and civilian circles, was a real break for me.

As October rolled around I handed command of No. 12 Group over to my successor, G/C John Easton, and prepared to head for Kingston. Our newly acquired furniture went into storage and our fairly new Hillman was loaded for a trip east through the United States.

29

A Very Senior Student

We set off at the beginning of October 1949, travelling from Vancouver through Portland to Boise and on to Salt Lake City, where we stopped to visit the Mormon Temple with its great pipe organ. From there we drove to Cheyenne, where we visited my old flying friend and boss from TCA days, Slim Lewis and his wife. Slim had "retired" to a large ranch. He had left TCA early in the war to be a test pilot at Boeing in Seattle. This was the last we saw Slim, who died in 1965.

From Cheyenne we drove to Denver to visit more friends from TCA days, Ted Larson and his wife Anne. Ted had returned to his original company, United Airlines, as a vice president. He passed away in 1966. From Denver it was on through Nebraska and Indiana, then north, crossing the border at Detroit and Windsor for the easy last leg to Kingston.

I checked in at the National Defence College in old Fort Frontenac, and we found a small, second-floor furnished apartment. The Commandant was Lieut.-Gen. Guy Simonds of Second World War fame. The directing staff consisted of G/C W.I. Clements (RCAF), Capt. A.H.G. Storrs (RCN), Col. A.F.B. Knight (Canadian Army), Morley Scott (Department of External Affairs) and Maj. R.G. Kingstone (Canadian Army). The 25 students were drawn from the senior levels of the three Canadian services and the RCMP, senior British and American officers, and top executives from government and industry.

Our studies, covering both the Canadian and world pictures, dealt with military, political, industrial and economic subjects as broadly seen from the highest levels. In addition to our normal work we each had to prepare a 5000-word thesis, to deliver as a lecture; my topic concerned Brazil.

We worked hard until December, when the directing staff and our student group took off in an RCAF North Star for a European tour. The captain was one of our best Air Transport Command pilots, S/L Benny Gruenwald. We flew from Rockcliffe to Goose Bay and on to Northolt, near London, arriving after 14 hours and 35 minutes.

We began a busy schedule, meeting political and industrial leaders, and visiting industrial plants. Between these assignments I was visited at the hotel by S/L J.H. Tudhope, whom I had not seen since before the war and who was now on staff at Canada House in London. Alas, he was another friend who was to die not much later.

On the 10th we left Northolt and flew to Orly Field, Paris. We stayed at the Hotel Scribe and our program in Paris was well handled by the Canadian Embassy, then headed by Gen. Georges Vanier, later the Governor General of Canada. We spent a day in Fontainebleau at Western Union Headquarters (the predecessor of NATO), commanded by my friend Field Marshal Montgomery.

Next morning the group was invited to breakfast at the apartment of G/C Carling-Kelly, then our air attaché in Paris, and his wife. Unfortunately the hotel failed to make all the early-morning telephone calls which had been arranged to ensure mass punctuality. A few of us were not called in time, and all was confusion as we clustered in the lobby, trying to get taxis in a hurry. To the rescue came Supt. George B. MacClellan of the RCMP, also a student on the course. He contacted his friends in the Paris Police Force. In minutes a police car was at the door, and we were driven across Paris at a terrific clip, with the horn going continuously. We were only a bit late, thanks to George and his friends. Subsequently George became Commissioner of the RCMP, and from 1967 to 1974 he was ombudsman for Alberta.

On December 13 we flew to Frankfurt for meetings with German and British leaders in the Occupied Zone. We gained an insight not only into the problems besetting the defeated Germans but also those of the occupation forces. A bus, driven by a former soldier still wearing part of his Wehrmacht uniform, took us along the Rhine Valley to Düsseldorf. We saw the ancient Rhine castles and stopped in Bonn to see the steps being taken to form a West German capital. We travelled around the Düsseldorf-Essen area looking at the terrific battering the region had received. Some of the people were still living in quarters scooped out under the rubble. Near Essen we toured the undamaged Krupp House.

On the 15th we took off from Düsseldorf for Northolt. Once back in Britain our busy schedules were on again. At Portsmouth we were taken out on a motor torpedo boat on manoeuvres. The sea was rough, the boats ran at high speed and we were all on deck. Our clothes were soaked and white with salt when we returned and it was tough getting everything clean again. In Glasgow we lunched with the Lord Provost, toured the Clyde shipbuilding areas, saw the boiler shops of Babcock and Wilcox, and drove through miles of Scottish industrial estates where factories were mushrooming. At Edinburgh we toured the castle as guests of the commandant. Then came the Royal Mile and other points of interest. The next day we took the train to London.

On December 22nd we took off in our North Star for Iceland on our journey home, landing at Keflavik after six hours. Here we were caught by bad weather until the 23rd, when it improved slightly. As we were all anxious to be home for Christmas, we pushed on. It was a rough IFR flight, with moderate to heavy icing. Iceland closed down again, so we couldn't turn back, and when we heard that Greenland was also below limits we knew we'd have to push through to Goose Bay.

Somewhere between Iceland and Greenland, I smelled smoke, then noticed a wisp of smoke near the cabin roof. I went to the cockpit, closed the door behind me, and found the front end permeated with smoke and the smell of burning rubber and metal. Smoke was coming through the instrument panel and the cockpit floorboards. Benny Gruenwald had given orders to remove a floor hatch into the baggage compartment below. An NCO crawled through the small opening and eventually extinguished a fire in the compartment. He reported that it was a large relay or other mechanism, fastened to the side of the compartment, which had caught fire. Strangely, all the electrical circuits seemed to function, so we were at a loss to know what the mechanism had been. I never did hear the exact cause. It was certainly a matter of life and death that the fire be extinguished, as we had no place to come down but in the seas below.

After eight and a half hours of battling weather we landed at Goose Bay, ate, refuelled and took off, arriving at Rockcliffe five and a quarter hours later. From there we were driven to Kingston, arriving late on the night of the 23rd. We were home for Christmas.

Soon we returned to our studies. Besides the twice-weekly lectures, delivered mostly by high-level lecturers from all over Canada and the world, we carried on our studies in syndicates. Also, we were now delivering the lectures we had prepared ourselves.

There were other trips. In February 1950 we visited New York City, where we attended sessions of the United Nations while our wives took in the shops; the evenings were devoted to theatre. In April we embarked upon a Canadian tour, with the West Coast and the High Arctic on the itinerary, and plenty of military and industrial sites to visit en route. This began on the 13th with a flight by North Star to Winnipeg. Over the next few days we flew to the tri-service base at Rivers, saw the oil fields at Leduc and the experimental station at Suffield, Alberta.

On the morning of the 17th I woke up with a very sore throat, but did not worry about it. We took off from Lethbridge, heading for Vancouver at high altitude. As we approached our destination we had to get down quickly as the aerodrome was on the verge of being closed in by weather. On the final part of the fast descent I developed the most excruciating pain in my ears and head, unlike anything I had ever experienced. By the time we landed I was in agony and almost deaf. Arthur Pidgeon, one of our students and an executive of the CBC International Service, had similar ear trouble. We finally wound up being treated in the Esquimalt Military Hospital. Apparently I had a damaged eardrum and severe middle-ear infection, and we both had trouble with our balance or equilibrium. We were taken off the flight, given penicillin and finally declared fit enough to travel back to Kingston by train.

Shortly after arriving an itching rash broke out over my body. It was apparently a penicillin reaction. The Kingston Military Hospital gave me adrenalin. Also, with Lin's help, I had to bathe myself in Calomel to ease the itching. Finally everything straightened out again, but I was left with a damaged eardrum.

The rest of our group returned to Kingston. I was sorry to have missed the trip, but it didn't matter much as I was familiar with the Arctic sites they had visited. Our studies continued until June 1950, when we graduated. It was, without doubt, one of the most interesting and valuable years of my life. We had been taught to think in broad, high-level terms, using insights into how the world and its people really operate, what actually makes it tick. We had studied the good and bad points of contemporary ideologies. My outlook on the world and its affairs was vastly broadened. Then, just as we graduated, a bombshell dropped on the world again. The Korean War had started.

30

Staff Officer, AFHQ and ATC

From the National Defence College I was posted to Air Force Headquarters as Director of Air Operations for the RCAF. With the Korean War just beginning, this was one of the real "hot seats" at AFHQ.

Lin and I found a second-storey apartment in an old house on Carling Avenue, east of Bank Street and near the Rideau Canal. Our Hillman had been traded in during our stay in Kingston and we now had a Morris Imperial Six, with a six-cylinder overhead camshaft engine. A good deal of my spare time was spent polishing and tuning it.

My new directorate was concerned with all RCAF non-training operations, including the Korean airlift. I reported to my friend Cars again, now A/C H.M. Carscallen, the Chief of Air Operations, and through him to A/V/M Frank Miller, the Air Member for Air Staff. My second-in-command was G/C Frank Ball. We had a number of branches such as Fighter, Air Transport, Tactical, and Radar.

We were involved in every aspect of air operations including the planning of the early warning radar lines. It was such a busy place that it was difficult to get away on trips. I did manage to get down to Chatham and Dartmouth and occasionally to Dorval to see what was happening in the field.

On July 25 six North Stars left Dorval for McChord Field, just outside Seattle, to start what became the famous Korean airlift. The aircraft were under the command of W/C C.H. "Cam" Mussells, a top officer who did a tremendous job in getting the operation under way. From McChord they were to fly to Anchorage, then to Shemya in the Aleutians, Tokyo, and sometimes Ashiya in southern Japan.

In December I had to fly to Montgomery, Alabama, to deliver a lecture at the USAF Staff College. I was not a very good speaker and intensely disliked lecturing, but managed to get the job done. My return flight, with a refuelling stop at Washington, involved some troublesome flying in icing conditions.

It was in January 1951 that we drove down to Grimsby, Ontario, to see a fruit farm which was for sale. Our friends, G/C F.J. Ewart (RCAF, retired) and his wife of Grimsby, thought the farm was a good buy. We saw the place with its 11 acres of fruit trees and decided to take it for future retirement purposes. Lin's father was already retired so he and his wife would move to the farm to look after it until we retired. From here on our holidays were always taken in the fall so that we could help pick fruit. When our retirement came in 1957, we moved Lin's parents into another house and moved onto the farm.

In February 1951 I went to McChord to see how the Korean airlift was progressing and to investigate the problems of various units en route. I borrowed a Dakota and crew from No. 412 Squadron. From Rockcliffe we flew to Lakehead, Winnipeg, Rivers, Edmonton, Vancouver and McChord. I spent a day with Cam Mussells and his 426 Squadron operation, which was functioning well, then made the return journey through Vancouver, Lethbridge, Winnipeg, Toronto and Rockcliffe. Total flying time for the inspection was 39 hours and 5 minutes, including 8 hours and 40 minutes of night flying.

In June 1951 I attended the annual Fire Power Demonstration at Eglin Air Force Base, Florida, a huge place where the use of new conventional weapons was demonstrated. I flew down via Washington and Montgomery on the 29th. The following day, with many other observers, I saw an impressive display of firing and bombing, but the most interesting of all was the sonic boom from a jet fighter. The "boom" was new at that time. A high-flying jet dived at our stand and pulled out, causing a sonic boom to hit the grandstand dead on. We were lifted about six inches from our seats by the wallop. Like some others, I was rendered almost stone deaf for a few hours. During the visit I saw Dave Callahan of Gilfillan Radio (the Ground Controlled Approach man) who was also attending the demonstration.

By October the pressures of the Air Operations job were making me tired and tense, so as a break I was transferred to Air Transport at Lachine to be Chief Staff Officer. The Air Officer Commanding was A/C R.C. Ripley. We had both grown up in Lethbridge and had served together in No. 11 (BR) Squadron at Dartmouth early in World War II. Rip was a good officer and friend who unfortunately was killed in a B-25 crash near Ottawa in 1957. It was good being back in my old Transport Command, in part because the job let me fly more. My health improved rapidly. Lin and I rented a house in Dorval, not far from headquarters in Lachine.

Air Transport Command was operating a scheduled military airlift to the United Kingdom and Europe, various scheduled operations in Canada, VIP flights for the Federal government, freight and personnel flights to our weather bases in the Arctic Islands and the Korean airlift, and carrying out aerial photography in northern Canada using a special squadron, No. 408. In addition to such routine duties we provided air transport and navigational escort required in flying Sabre fighter squadrons to the United Kingdom and Europe, through Greenland and Iceland (Operation Leap Frog). All in all, we seldom had much time to spare.

During my time as Director of Air Operations at AFHQ I had teamed up with my old boss from the bush days, Punch Dickins of de Havilland Aircraft, to promote the idea of the RCAF buying two Comet jet transports. We succeeded in convincing A/C Carscallen and A/V/M Miller, and in January 1952 we received instructions at ATC Headquarters to despatch a team to England to plan the back-up of spares, pilot training, other aircrew instruction, and the training for ground crews. All this had to be done before a totally new and different type of aircraft like the Comet could be operated successfully. We were the first air force in the world to operate jet transports.

Punch flew to England ahead of us to get things started. The supply and technical teams followed, then the operational team composed of W/C W.H. "Bill" Lewis, our chief engineer, S/L J.A. "Andy" Anderson and myself, flying aboard a TCA North Star.

On January 8, 1952, the three of us, along with Punch, made our first flight in a Comet — one used by de Havilland as a test aircraft — and with a test pilot at the controls. This Comet had few seats, most of the space being occupied by test equipment. We flew for 2 hours and 15 minutes, doing a series of relighting tests at 28,500 feet. This meant shutting down one engine, letting it get thoroughly cold, then firing it into operation again. It worked very well, but we were startled on each relight, which was accompanied by a great boom and a sheet of flame from the tailpipe. Nevertheless, we were much impressed by the Comet. We arranged for the manufacturers to train our first pilots, navigators, wireless operators and flight engineers on the Comet. Afterwards they would have further training in Canada.

On the 16th we left England by TCA North Star via Prestwick, Iceland, Gander and Dorval. I was soon on the move again, first to Washington, on air transport business, followed by a trip to Japan and Korea with our airlift to get a first-hand look at the operation, now being presided over by W/C J.K.F. MacDonald, who had succeeded W/C Mussells as CO of No. 426 Squadron.

I left Dorval on February 3, 1952, aboard one of our scheduled North Stars, flying to Edmonton with one refuelling stop. Next day I had work to do in Edmonton with the local transport squadron. I also accepted a flight in a new Bristol helicopter which had impressive performance. I had now accepted completely the "whirley-bird" as a valuable piece of equipment.

From Edmonton I flew to Vancouver and on to McChord, arriving late on the 4th. The next day we took off, heavily loaded, for Elmendorf airfield at Anchorage with F/O Harrison as captain. After a meal and refuelling at Elmendorf we left at night, flying down the Aleutian chain to Shemya, at the western end. Flying time from Edmonton was 19 hours and 10 minutes.

The next leg of the flight, with F/L Rowly Lloyd as captain, was a night departure on the 6th. There had been heavy blowing snow for some time and on take-off we hit some drifts on the runway, which made the moment a dicey one. I sent back a strongly worded radio message to the base which led the Americans to improve their runway clearance procedure. We finally arrived in Tokyo after a flight of almost 11 hours.

In Tokyo I stayed at the Marinuchi Hotel, used only by the military and controlled by the Australian armed forces. Our RCAF air attaché in Tokyo was G/C R.W. "Buck" McNair, the famous fighter pilot. Buck took me to the various military headquarters, introducing me to senior American commanders and arranging several briefings. Permission was granted to fly to Korea to see operations there. Buck organized all the details for the Korean visit. From Tokyo I was to be accompanied by W/C E.G. "Irish" Ireland, another famous RCAF fighter pilot, visiting Korea to study fighter operations there.

On February 12 we went to Yokahama, to tour some of the shipbuilding industries, then to Tachikawa airport where we boarded a USAF C-54, with Maj. Reynolds at the controls. After two hours of flying, mostly over mountainous country, we landed at Kohama, where we discharged some of our load, took off again, and arrived at Ashiya, in southern Japan, two hours further on. This was a big air base almost directly across the water from the tip of Korea.

Irish and I were given quarters in the commanding officer's house, where we were fed choice Japanese foods prepared by the colonel's Japanese servants. We were also shown around the area. The USAF was operating a big air supply operation from Ashiya using C-119s. They carried everything from food to field guns and ammunition, which were paradropped on pallets. They specialized in supply dropping in the Korean mountains to troops cut off or surrounded, and had lots of business. At this time they were having serious trouble with the C-119's propellers, as we were in Canada with our C-119s.

On the 14th we took off with Capt. Warr and flew to Osaka, then across the sea and Korean mountains to Seoul, South Korea. The city was terribly smashed up, in places nothing but rubble. The airport was a shambles, complete with burned hangars. Heavy steel matting constituted the runways on which we landed with a noisy clatter. We were met by Col. Alan Towne, USAF, who had formerly been with United

Airlines. He was amiable and arranged beds and meals for Irish and me in his Korean home. The house had a stone hot-water boiler at the back and the heated water from the boiler flowed through stone ducts under the floors. During cold weather the floors and lower walls were warm. It was quite comfortable when there was snow on the ground outside.

Irish was later taken off to see the fighter aerodromes, and I went north across the Han River to the battle zones, north of which the supply drops were conducted. In Seoul we met senior American commanders and attended some of their battle briefings. Being mountainous, Korea was tough country in which to fight, and since North and South Koreans looked pretty much alike there were difficulties distinguishing friend and foe.

Many South Koreans had lost their children as well as their homes in the war. When the Communists were driven out of Seuol they took many South Korean children with them, never to return. We were told that during the Communist occupation the children had been drilled in Communist ideology. Some had been trained as arsonists to destroy buildings. Just after the Communists were driven out, some prominent buildings were burned. At first the populace thought the Americans were to blame, but learned later that their own children were involved, obeying the instructions of their departed Communist masters.

On the night of the 16th we boarded a C-54 loaded with lightly wounded soldiers and some who were due for rest and relaxation. The weather was poor, but we climbed above the mountains on instruments and headed for Ashiya. Somewhere well south in Korea, still flying high and blind, there was a bump and a flash of fire on the starboard side. The flame went out fairly fast when the crew used the fire extinguishers. One of the engines had thrown a connecting rod and momentarily caught fire. We managed to struggle into Ashiya despite our load.

The next day, the 17th, we took another aircraft first to Seoul, and then to Tachikawa, near Tokyo — a roundabout way to our destination. Back in Tokyo we gave Buck McNair an account of our trip, bought presents on Ginza Street to take home, and were ready to go. Departure was arranged in one of our North Stars which was returning to Canada on the night of the 19th.

We boarded at Haneda Airport, Tokyo, with F/L Stuart as our pilot. Wet snow was falling. I spoke to Stuart about clearing the wings off but he felt it would blow away on take-off. I wondered, but did not press the matter. I should have. We rolled down the runway towards the bay but our acceleration was slow. The snow was not blowing off the wings as fast as expected. Finally the pilot had to yank the aircraft off the ground, but we collided with a low sea wall at the end of the runway. It seriously damaged the nosewheel, which refused to retract and thus prevented closing the nosewheel doors. The aircraft climbed slowly for a long time through the darkness until the last of the snow was gone from the wings. There was an icy gale blowing through the cockpit from the open nosewheel doors, and the aircraft cruised more slowly than usual.

It was a long, cold night across the Pacific as we went up to the cockpit to spell the pilots off. It was too cold to sit there for long. Eventually, after 9 hours and 15 minutes of night and cold, the runway of Shemya appeared. F/L Stuart landed safely with the damaged wheel, but the aircraft had to be repaired at Shemya. We had to wait for the next flight going east.

On February 21 we boarded another North Star, piloted by F/L Roane, and flew to Anchorage. After refuelling and eating we carried on to McChord, and thence to Vancouver and the flight home to Dorval, where we arrived on the 23rd.

It was wonderful to be home to catch up on Lin's cooking and on my work at Transport Command headquarters. The latter included the annual inspection trip at the end of March and beginning of April to Goose Bay, now commanded by G/C L.J. Birchall, the "saviour of Ceylon" in 1942 and long a Japanese prisoner of war.

On April 19 I set out to inspect our Arctic air bases and weather stations, which included viewing the annual resupply of the bases by our transport squadrons. We left Dorval in a fully loaded North Star and flew direct to Fort Churchill. After refuelling we flew north over the Barren Lands, crossing the Arctic coast, passing King William Land, the Boothia Peninsula, Somerset Island, and then to Cornwallis Island. Our base at Resolute Bay was on the southern end of Cornwallis. We were now well north of the north magnetic pole.

Resolute Bay was at the centre of Arctic operations. It had one good long runway and facilities for Arctic living. The water and sewage systems were masterpieces of ingenuity. Buildings were linked by utilidors — corridors containing the heating and water pipes, electrical wiring, etc., through which personnel passed from one building to another. Utilidors were particularly valuable during Arctic blizzards, when travel outside was hazardous. Every precaution was taken to guard against fire, one of man's greatest enemies in these latitudes. All paint, for example, was fire-resistant.

The resupply operation was then in progress. Great piles of supplies, including 45-gallon drums of heating oil, were concentrated at Resolute and moved by air to the surrounding weather bases. The supplies were brought into Resolute in summer by ice-breaking ships which sometimes, if ice conditions allowed, were able to get to one or two of the other bases.

The next morning, April 20, we were to take a load to Alert, our most northerly weather station on the northern tip of Ellesmere Island, a little over 400 miles from the north pole. We headed north over Devon Island, past Axel Heiberg Island and along Ellesmere Island. The mountain ranges on Ellesmere in places are almost as high as the Rockies, requiring plenty of altitude. They are known as the British Empire range, the United States range, and the Victoria and Albert range and are an awe-inspiring sight on a clear, cold day. The valleys between the peaks are filled to a great depth with hard-packed snow.

We landed on the snowy gravel strip at Alert and visited the campsite during unloading. The camp was small but well equipped, and the personnel there were getting on with the job quite well. From Alert we flew down Ellesmere to the base at Eureka but couldn't get in due to weather. From here we swung in an easterly direction to Greenland, where we landed at the big USAF base at Thule. Thule was a good alternate landing field for our Arctic flights in case of bad weather at Resolute, which in turn was used as a USAF alternate when Thule was "socked in." It was comforting to have such facilities in this area.

After lunching at Thule we returned to Resolute in 2 hours and 50 minutes. The next day I wanted to visit Isachsen weather base on Ellef Ringnes Island, so our aircraft was loaded with supplies for that base. Unfortunately the weather closed in just before we arrived. We made two attempts to land but the visibility was too poor and we had to return to Resolute. Satisfied with my inspection, I returned to Dorval on the 21st and 22nd, stopping at Churchill and then Trenton where I checked our air transport training school.

During this year, 1952, I was saddened to hear of the death of "Wop" May. After our bush flying days at Fort McMurray in the early '30s I had not seen Wop again until near the end of the war. He had come to Ottawa on business and had visited us for an evening. We had a lot to catch up on. Later I visited Wop's Air Observation School which he was running in Edmonton for Canadian Pacific Airlines. He put on a demonstration of parajumping by two men, including a Sgt. Hargraves. The jumping took place over bush and the men dropped right into the trees with their special gear. I was impressed with this demonstration and the search and rescue possibilities of pararescue. I reported this to the Chief of Air Staff in Ottawa. This was the beginning of today's pararescue operation, pioneered by Wop May and his men. That demonstration was the last time I saw Wop.

Air Transport Command Headquarters continued to hum with activity, and I also made a number of brief visits to places in eastern Canada — Goose Bay, Fort Chimo and Frobisher Bay in October, Resolute Bay again in November. In the spring of 1953 we had another visit by Field Marshal Montgomery. I was one of the party greeting him.

Along with others, in June I was presented with the Queen's Coronation Medal by my chief, A/C Ripley. That month I also managed to get a trip west in one of the Comets which were being fitted into our organization. The pilots were W/C Howie Morrison and S/L Stew Olsen. Both had completed training courses in the United Kingdom. We were anxious to fly a Comet across Canada to bring to light any problems in air traffic control between jet and piston-engined airliners. Many problems did come to light, such as the great difference in cruising altitudes of the two types and also the matter of jets descending and ascending through piston-engined airways.

On June 8 we took off in Comet 5301 with Morrison at the controls, climbing to high altitude and then down again into Toronto, with a flying time of 1 hour and 20 minutes. Our altitudes and speeds were revolutionary at this time; wherever we landed crowds assembled to see the aircraft. From Toronto Stew Olsen took over and after reaching a cruising altitude of 40,000 feet flew to Winnipeg in 3 hours.

On the 9th we left Winnipeg for Saskatoon and Edmonton, where I had work to do so the Comet went on. The flights in the Comet had shown the great potential of the jet airliner. After completing three days of work in Edmonton I flew to Winnipeg in a C-119 and then to Montreal in a North Star.

In October 1953 Rip Ripley was posted to Fontainebleau in France and I was made Officer Commanding, Air Transport Command. I was sorry to see Rip go, as he was a good commander and a fine friend. I was busier than ever, travelling widely by air as I looked into the various units and operations, including our internal flights, services to Europe and the Far East, and the aerial photography of the north.

In January 1954 I attended a two-day conference of Air Officers Commanding, held in Ottawa and chaired by A/M Roy Slemon, who had replaced the retiring A/M Curtis as Chief of the Air Staff. At the conference each commander outlined the requirements of his command for the coming year and justified his needs. Present was A/M John Plant, now senior RCAF officer in Europe.

On an April trip to the Arctic islands I was able to confirm that the resupply operations were proceeding well despite grim weather. Then in June 1954 the Korean airlift operation was terminated, the 599th and last round trip being completed by W/C Bill Lupton. Since its inception in July 1950 the operation had been operated without injury to a single passenger or loss of a pound of cargo.

Our two Comets had been used simulating enemy jet bombers by flying penetrations into various areas of Canada, alerting and testing our air defence installations and crews. In addition they were used as VIP transports. We had had few troubles with them, but two were lost overseas by commercial operators in tragic crashes. They were coming apart in the air following sudden decompression caused by the fuselage skin cracking around the square cabin windows. After long testing in the United Kingdom this fault came to light, but it was found that it did not occur in airframes with under 3000 hours. Our Comets were quite safe at this point so we continued to operate them. Later, however, they were returned to England where the fuselages were modified. This done, they returned for service in Canada.

In September 1954 I heard that I was to be posted to Toronto as Group Commander of No. 2 Group, with A/C (later A/V/M) Carscallen taking over Air Transport Command. At the end of the month we had a handing-over ceremony. Then Lin and I packed and headed for Toronto.

31

Recessional

Group Captain (later A/C) Val Patriarche relinquished the command of No. 2 Group on posting to Ottawa and I took over. Val, his wife, Betty, and family moved out of the Group Commander's quarters, known locally as the "Dog House," and Lin and I moved in.

No. 2 Group was a mixture of Regular and Reserve Forces. It consisted of an Operations Wing and a Technical Wing. The Operations Wing was composed of two flying squadrons, Nos. 400 and 411, and an Aircraft Control and Warning Squadron. The Technical comprised a ground training squadron, a radar squadron and a large medical unit. The Operations Wing was commanded initially by G/C George Gooderham, then later by G/C R.C.A. "Bunt" Waddell, with W/C Chuck Darrow as second-in-command. The two flying units were commanded by S/Ls W.A. Curtis (the son of A/M Curtis) and W.H. "Bill" Draper; the Technical Wing by G/C O.B. Coumans, with W/C Earl Mann as his second-in-command. Regular Force support personnel were placed throughout the units to assist.

Lin and I had barely settled in the house, and I was just beginning to find out what the group looked like, when Hurricane Hazel hit Toronto. Heavy cloudbursts flooded the countryside. Rivers ran wild, washing away houses and drowning dozens of people. I alerted the group and sent our men out to help wherever needed. With Bill Draper in a Harvard I went aloft to reconnoitre the destruction from Holland Marsh to Lake Ontario. It gave us some idea of where to assist, mainly with our men along the river banks.

Throughout the winter of 1954-55 operations went on normally, the squadrons flying Vampires. Later they re-equipped with Sabres, the hottest aircraft ever flown by the Auxiliary squadrons. I had been having increasing difficulties from sinus trouble and my damaged eardrum. Higher altitudes bothered me somewhat, particularly in unpressurized aircraft. In March 1955 I began the High Altitude Indoctrination Course at the Institute of Aviation Medicine. I got through most of it but finally had to be taken from the high altitude chamber because of sinus pains. It was apparent that high altitudes, at least without pressurization, were no longer for me.

Also in March, I was despatched to Camp Borden to attend the Senior Officer's Course in Atomic, Biological and Chemical Warfare. The school was being run by Lieut.-Col. V.R. Schjelderup (Canadian Army) and the students included a host of old friends — A/C Carscallen, G/C Baxter Richer, Col. Jerry Charlebois and others. I was particularly interested in the atomic portion of the course, and have since maintained a fascination with the atom and the universe. On my return to the group I found all was running normally. I began regularly flying the twin-engined Expeditor which enabled me to keep my hand in at flying without forcing high altitudes upon me. One of my trips took me to Sudbury in March 1956, where I acted as president of a Court of Inquiry into a matter at RCAF Station Falconbridge. There I visited an old friend from No. 13 (OT) Squadron days, W/C Chuck Austin, now retired from Austin Airways, as well as Frank Russell, Austin Airways' chief engineer.

The Auxiliary had now become proficient on Sabres. I was proud of their efforts as they were mainly Reserve personnel who had little time to spare for RCAF duties. Regular Force support personnel had done a splendid job training and assisting the Reservists. Among those unsung helpers were S/Ls Pattinson, Bishop and Dobson, F/Ls Milne and Gilland, and F/O Al Rokeby.

I learned that I would be retired on reaching the maximum age for my rank, on June 19, 1957. I could take my retirement leave before or after that date, so decided to take it before. My work would finish, therefore, on February 13, 1957, although I would remain on full pay until June 19, after which I would be placed on the Retired List and on pension.

With W/C Joe Mirabelli, the Commanding Officer of RCAF Station Toronto, I flew my Expeditor to St. Hubert in the first week of February, where the AOC, Air Defence Command (and thus my boss for this last period) A/V/M L.E. Wray, put on a Mess Dinner and presented me with a special plaque. With terrific hangovers, Joe and I flew back to Toronto the next day.

On the night of February 12, Joe Mirabelli put on a Mess Dinner at Station Toronto where I made my final farewell speech. Joe had invited many of my old friends, and of course Lin, to be present. The guest list was a "Who's Who" of Canadian flying "greats" with whom it had been my privilege to work — A/M W.A. Curtis, A/V/M George E. Brookes (the first commander of No. 6 Canadian Bomber Group, 1943), Jack Austin of Austin Airways, Frank Young of TCA, A/C Alex Gilchrist from Air Defence Command, and others. Many beautiful gifts were presented by Station Toronto and No. 2 Group to Lin and me. Finally it was all finished. We had come to the end of a long trail of unforgettable friends and flying.

We moved from Toronto to our small fruit farm at Grimsby. At the request of chairman Frank Young I acted as director of operations at the Canadian International Air Show, Canadian National Exhibition, for nine years. Eventually I found myself getting out of date in some respects, so I passed the job on to more qualified people in 1966.

I am glad that my career in aviation happened during the years described here. They were fascinating beyond words. I think Lin was quietly relieved when it was all over. She had flown with me many times, lived in so many places, good and bad, and waited for me to return from so many trips, always supporting me in my endeavours. Without her I could not have made it. I was determined to fly, and I did fly, but now I was home to stay.

Appendix

"The Beginning"

My interest in the world of the atom commenced during the atomic part of a military training course I took in 1955. Having always been interested in the study of the Universe, it seemed a natural step to combine the two studies, inasmuch as we now know that the Universe is made up of atoms.

We know that energy and matter are interchangeable and that all matter is composed of atoms, therefore it would appear that the atom, and its sub-divisions, are the building blocks of the Universe. All matter, whether it be rocks, trees, air, water, soil, plants, birds, animals, etc., and even our own human bodies, is composed entirely of atoms, the various combinations of which determine what particular shape and type of end-product will follow.

We have learned that each particular atom will retain its identity unless it is in collision with other atoms or their sub-divisions. When such collisions occur, whether naturally or artificially, different types of atoms are the outcome. In some cases the atom (or matter) is changed to energy by the collision, such as in atomic or hydrogen explosions. This process is going on in our sun and active stars.

The beginning of the Universe is thought by some to have resulted from a gigantic explosion, by others to have been created in a few days by a Supreme Being, and another theory is that there never was a beginning, that the Universe has always existed.

I find all these theories difficult to accept without at least some modification. In the case of the "Big Bang" it is clear that a huge explosion would require sufficient matter, gases, etc., to be present before the explosion. Where did they come from? They had to be there *first.* Also it appears that there is a process of aging, of birth and death, of a beginning and an ending, in most things, if not all. If there was a beginning there must have been a time when there was nothing. If there was nothing there must have been a great void with no form of matter or energy in it. This is very difficult to visualize, but if there was a beginning it must have been so.

Since it is accepted that the atom, with its subdivisions, is the basic building block of the Universe, it should follow that the creation of the first atoms must have been the beginning of the first matter in the void. It should also be borne in mind that to have one or more atoms appear in the void, at that time, is scientifically impossible. In addition, once the first atoms did appear in the infinite void, the possibility of collisions in any, or even sufficient quantity, is also scientifically improbable or impossible.

It is clear, therefore, that if the Universe had a beginning, as I believe it did, there are two scientifically insurmountable obstacles. First, the appearance of the atoms, and second, sufficient collisions to start a process of growth. Therefore if the idea of a beginning is accepted, the creation of the first atoms from absolutely nothing, and the guidance necessary for the first period of collisions, must be supernatural. There is no other way. This, then, is the work of a Supreme Being.

At this point it is important to fully realize that the planning and creation of a number of complicated atoms with their subdivisions, from nothing, because nothing existed, is beyond human imagination. Yet if there was a beginning, this must have occurred. Once the process of creation of atoms got properly started there would be bodies of matter, planets, etc., coming into being over a very long time.

Today we accept the concept of Evolution, or the natural order. We can see it around us, in our animals, plants, trees, and particularly in the human race itself, which was given a special place within the Plan. There seems to be no question but that the natural order is the process which is in operation. The natural order, evolution, must therefore be the Plan of things.

If we accept the thought that there was a beginning from nothing, then the planning and operation of the first atoms was Creation. If that was so, then, since we know that we are part of a natural order of progression, then this must be intentional. It must be the Plan of the Supreme Being, the Creator of the Atoms.

The Plan of the Creator, now in action, has produced over a very long time a magnificent, precise Universe, which continues to grow. Eventually life begins on Earth, and possibly elsewhere, and the Natural Order Plan controls the process, except where the Creator decides to alter it; otherwise He lets the Plan continue.

In the Plan man runs his earthly affairs, including his own good and bad actions, his wars, his greed, selfishness and so on. The Creator does not interfere unless He is unselfishly and sincerely asked to do so, and if He is in agreement He may decide to grant the request. If He does so He may carry out His work in ways not clearly obvious to man: i.e., cure sickness through the skilled hands and knowledge of physicians and scientists; or help people with problems of faith or the soul through the hands of theologians; or end wars or provide shelter and food through the works of statesmen and politicians.

In the end, however, if man eventually loses control of the situation, the Creator will assume total control and will rectify all problems.

Z.L. Leigh
With the assistance of Mr. Dale Farnham
August, 1969
May 19, 1984

Index